INFLATION AND QUANTUM COSMOLOGY

INFLATION AND QUANTUM COSMOLOGY

A. D. Linde

Lebedev Physical Institute
Moscow, USSR

Edited and with a Foreword by

Robert Brandenberger

Department of Physics
Brown University
Providence, Rhode Island

ACADEMIC PRESS, INC.
Harcourt Brace Jovanovich, Publishers
Boston San Diego New York
Berkeley London Sydney
Tokyo Toronto

This book is printed on acid-free paper. ♾

ACADEMIC PRESS, INC.
1250 Sixth Avenue, San Diego, CA 92101

United Kingdom Edition published by
ACADEMIC PRESS LIMITED
24-28 Oval Road, London NW1 7DX

This work is based on the following articles, reprinted with permission from their respective journals.

Chapter 1

A. Linde, "Particle Physics and Inflationary Cosmology," *Physics Today* **40**, 61-68 (1987).

Reprinted with permission from *Physics Today*, September 1987, pp. 61-68, © 1987 American Institute of Physics.

Chapter 2

L. Kofman and A. Linde, "Generation of Density Perturbations in Inflationary Cosmology," *Nuclear Physics B* **282,** 555-588 (1987).

A. Linde, "Inflation and Axion Cosmology," *Physics Letters B* **201**, 437 (1988).

A. Linde, L. Kofman, and J. Einasto, in letter to *Nature* **326**, 47 (1987).

Chapter 3

A. Goncharov, A. Linde, and V. Mukhanov, "The Global Structure of the Inflationary Universe," *International Journal of Modern Physics* **A2**, 561 (1987).

A. Linde, "Chaotic Inflation with Constrained Fields," *Physics Letters B* **202**, 194 (1988).

A. Linde and M. Zelnikov, "Inflationary Universe with Fluctuating Dimension," *Physics Letters B* **215**, 59 (1988).

A. Linde, "Life after Inflation," *Physics Letters B* **211**, 29 (1988).

A. Linde, "Universe Multiplication and the Cosmological Constant Problem," *Physics Letters B* **200**, 272 (1988).

"Particle Physics and Cosmology." Plenary talk given by A. Linde at the 24th International Conference on High Energy Physics in Munich, August 1988.

A. Linde, "Life after Inflation and the Cosmological Constant Problem," CERN preprint TH5389/89, to appear in *Physics Letters B* (1989).

LIBRARY OF CONGRESS CATALOG CARD NUMBER: 89-82199

ISBN 0-12-450145-1

Printed in the United States of America
90 91 92 93 9 8 7 6 5 4 3 2 1

Foreword

With the invention of the theory of the inflationary universe in the year 1980 cosmology entered a new area. Whereas the previous theory, the hot big bang model, successfully predicted the existence of the 2.7^o K microwave background radiation which permeates space and could explain the formation of light elements in the early stages of the universe, it left many questions unanswered. What is the origin of the observed homogeneity and isotropy of the universe; what is the source of the energy density perturbations which eventually lead to galaxies, clusters of galaxies, and structures on even larger scales?

The inflationary universe is the first theory to provide good answers to these questions based on a causal microscopic theory of matter coupled to general relativity as a theory of space and time. The crucial step was the realization that in the very early universe at high temperatures and energies a classical description of matter breaks down and must be replaced by a description in terms of quantum fields, known from accelerator experiments to provide a better description at high energies.

When considering quantum field theory in the very early universe, one discovers that in accordance to many theories of elementary particles the equation of state is likely to go through a phase with negative pressure p. In many theories at the very early stages of the evolution of the universe $p = -\rho$ where ρ is the energy density. This leads to exponential expansion of the universe which has the potential of wiping out any preexisting inhomogeneities and anisotropies. In 1982 it was discovered that quantum fluctuations during the period of exponential expansion generate classical energy density perturbations which look like small amplitude plane waves of all wavelengths superimposed with a spectrum which is in good agreement with the requirements for successful galaxy formation. In the years between 1982 and 1988 inflation has become part of the "standard theory of galaxy formation."

In October 1987 Professor Andrei Linde delivered a series of five "Morris Loeb Lectures" at Harvard University on inflation and quantum cosmology. Professor Linde has since the early 1980's been one of the two or three leading cosmologists who have contributed to the development of the inflationary universe scenario. In the 1970's he was one of the first to realize the important effects of finite temperature in quantum field theory and its profound impacts on cosmology. He published many articles discussing for the first time phase transitions in the early universe induced by finite temperature effects. Shortly after Guth formulated the inflationary universe scenario, Linde was the first to propose a model in which it could successfully be realized (the "new inflationary Universe," which was independently proposed also by Andreas Albrecht and Paul Steinhardt). Soon after that, Professor Linde proposed an improved model, the "chaotic inflation scenario" which has remained until now the most successful realization of inflation.

Since 1983 Professor Linde has almost single-handedly developed the inflationary universe scenario in great detail. A lot of this later work has been unjustly neglected in Western countries where old realizations of inflation have been canonized far too long.

This book is based on Professor Linde's Harvard lectures. The first chapter gives a brief but complete overview of the inflationary universe scenario. The problems of the standard big bang model are discussed, followed by a discussion of the main ideas of inflation. Linde then discusses chaotic inflation and its predictions for the structure of the universe on scales much larger than are visible to us.

As mentioned above, the inflationary universe is the first theory which by a causal mechanism predicts energy density fluctuations which can lead to galaxy formation. The second chapter of this book discusses in detail the mechanism by which these perturbations are produced. It also outlines the different kinds of final perturbation spectra which can be generated in inflationary universe models. Until recently, it was generally believed that the prediction is unique: a scale invariant spectrum of adiabatic perturbations. However, Professor Linde shows that in all but the simplest models also isothermal predictions are generated.

The third chapter of the book discusses in detail the most recent developments in inflationary universe cosmology. The main theme is stochastic inflation. Stochastic inflation leads to a picture of the universe on very large scales, scales far beyond our present horizon, which is very different from what people imagined in the early days of inflation. Quantum fluctuations cause the universe to fragment into an ever increasing number of mini-universes, most of which inflate again and in turn begin to fragment. Thus, on very large scales, the universe is very inhomogeneous, consisting of many bubbles in which the physical properties are very different. Coupling constants, microphysical laws and even the dimensionalty of space-time may vary from one domain to the next. These questions concerning the large-scale structure of the universe are closely tied to the mystery of the vanishing of the cosmological constant, as Professor Linde points out in the later sections of this chapter. Linde relates his approach to other approaches to the cosmological constant problem based on various formulations of quantum cosmology which have become popular recently.

This book is the first self-contained but nevertheless reasonably brief exposition of the recent developments in inflationary universe cosmology. In an appendix Professor Linde summarizes his view of inflationary cosmology in a sequence of cartoons which are extremely instructive (and also lots of fun to look at). In Professor Linde's words, it is important to make complicated material look simple enough such that even academics can understand it.

Chapter 1 is a slight modification of a survey article by Linde entitled *Particle Physics and Inflationary Cosmology* which appeared in *Physics Today*, 40, p. 61, (Sept. 1987).

Chapter 2 is based on an article by L. Kofman and A. Linde entitled *Generation of Density Perturbations in Inflationary Cosmology* which was published in *Nuclear Physics* B282, p. 555 - 588 (1987). Some portions are taken from the paper *Inflation and Axion Cosmology* in *Physics Letters* 201B, p. 437 (1988), and a small segment from a Letter to Nature which A. Linde wrote in collaboration with L. Kofman and J. Einasto (*Nature* 326, p. 47 (1987)).

Chapter 3 combines material from several articles:
The Global Structure of the Inflationary Universe by A. Goncharov, A. Linde and V. Mukhanov, in *Int. J. of Modern Physics* A2, 561 (1987).
Chaotic Inflation with Constrained Fields by A. Linde, in *Physics Letters* 202B, 194 (1988).
Inflationary Universe with Fluctuating Dimension by A. Linde and M. Zelnikov, in *Physics Letters* 215B, 59 (1988).
Life after Inflation by A. Linde, in *Physics Letters* 211B, 29 (1988).
Universe Multiplication and the Cosmological Constant Problem by A. Linde, in *Physics Letters* 200B, 272 (1988).
Particle Physics and Cosmology, a plenary talk by A. Linde given at the 24'th International Conference on High Energy Physics in Munich (August 1988).
Life after Inflation and the Cosmological Constant Problem, CERN preprint TH5389/89 by A. Linde, to appear in *Phys. Lett.* B (1989).

We would like to express our gratitude to Suzanne Guillet for her superb job of typing the manuscript.

Robert H. Brandenberger
Professor of Physics

CONTENTS

Chapter 1

PARTICLE PHYSICS AND INFLATIONARY COSMOLOGY

With the invention of unified theories of strong, weak, electromagnetic and gravitational interactions, elementary particle physics has entered a very interesting and unusual stage of its development. The end of the 1960s saw the introduction of the Glashow-Weinberg-Salam unification of the weak and electromagnetic interactions. In 1974 came the grand unified theories of the strong, weak and electromagnetic interactions. Two years later we had supergravity, giving us the first hope of unifying all fundamental interactions, including gravitation. The beginning of the 1980s witnessed a renewal of interest in the Kaluza-Klein theories and supergravity in higher-dimensional space-time. Nowadays superstring theory is the leading candidate for the role of "theory of everything."

The 1983 discovery of the W and Z weak vector bosons predicted by the Glashow-Weinberg-Salam theory has shown that the unified theories have made a very good start. Optimists are full of enthusiasm, predicting that at this rate we will soon reach our final goal. It would be highly desirable, however, to verify experimentally that we are still going in the right direction. Unfortunately, this appears to be an extremely difficult task.

The energy scale at which the unified nature of all four fundamental interactions is expected to become manifest is not very different from the Planck mass M_P, about 10^{-5} grams, where quantum gravity effects become important. (The Planck mass is that mass for which the Compton wavelength l_P, about 10^{-33} cm, equals the Schwarzschild radius.) Its rest energy $M_P c^2$, about 10^{19} GeV, corresponds to the kinetic energy of a small airplane. By contrast, the 80-km-circumference Superconducting Super Collider the Americans hope to build in the near future will accelerate particles up to 10^4 GeV. The largest accelerator ring one could build on Earth, with a circumference of 40,000 km could not accelerate particles beyond about 10^8 GeV, a center-of-mass energy occasionally to be seen in cosmic-ray collisions. But this still leaves us 12 orders of magnitude short of the energy necessary for a direct test of the unified theories. Of course there are some indirect tests, such as the searches for proton decay and for supersymmetric partners of ordinary particles. But trying to get a correct theory of all fundamental interactions with only such low-energy experiments is like looking for the correct unified electroweak theory by studying nothing but radio waves.

1. The Universe as Accelerator

The only accelerator that could ever produce particles energetic enough for a direct testing of the unified theories of all fundamental interactions is our universe itself. The Big Bang scenario as it stood ten years ago [1], which I will call the hot-universe theory, asserts that the universe was born at some moment $t = 0$ about 15 billions years ago, in a state of infinitely high temperature T and infinitely large density ρ. Of course one cannot really speak of classical space-time for the earliest moments, when kT/c^2 was greater than the Planck mass, and ρ exceeded $M_P/{l_P}^3$, making quantum fluctuations of the metric predominant. (We will use a convenient unit system that sets k, c and h all equal to 1, so that l_P equals $1/M_P$ and the Planck density is ${M_P}^4$, roughly 10^{94} g/cm^3.) It is just at such times, when the average particle energy exceeded M_P, that the unity of all four fundamental interactions would have been manifest.

With the rapid expansion of the universe, the average energy of particles, given by the temperature, decreases rapidly, and the universe becomes cold. The temperature falls as the reciprocal of R, the scale factor, or "radius," of the universe. This means that particle interactions at extremely large energies can have occurred only at the very early stages of universal evolution. One might think it very difficult to extract useful and reliable information from the unique experiment carried out about 10^{10} years ago. Thus it came as a great surprise to those who study elementary particles that the investigation of physical processes at the very early stages of the universe can rule out most of the existing unified theories.

For example, all the grand unified theories predicted the existence of superheavy stable particles carrying magnetic charge: magnetic monopoles. These objects have a typical mass 10^{16} times that of the proton. According to the standard hot-universe theory, monopoles should appear at the very early stages of the universe, and they should now be as abundant as protons. In that case the mean density of matter in the universe would be about 15 orders of magnitude higher than its present value of about 10^{-29} g/cm^3.

Originally there was some hope that this problem would disappear when more complicated theories were considered.

It turned out, however, that the problems became even more complicated in the new theories. For example, according to the models based on one version of supergravity, the universe should contain not only monopoles but also heavy gravitinos (superpartners of the graviton with spin $\frac{3}{2}$) and oscillating classical "Polonyi" fields. In most of these theories the predicted abundance of gravitinos (or their decay products) contradicts cosmological data by about 10 orders of magnitude, and the typical energy density stored in the Polonyi fields contradicts the data by 15 orders of magnitude. Most of the higher-dimensional Kaluza-Klein theories considered in the early 1980's predict the present vacuum energy density to be on the order of ${M_P}^4$, too large for the observational data by 125 orders of magnitude. The situation with the currently fashionable superstring theories seemed, at first glance, to be somewhat better. To be fair, however, one must say that no consistent cosmological model based on superstrings has been suggested thus far.

We see that it is not difficult to obtain strong cosmological constraints on the unified particle theories. On the contrary, there is a problem as to whether it is possible to reconcile elementary particle theory and cosmology. To answer this question it is necessary to check whether the standard hot-universe theory is as good as it is seemed at first to be, and whether it is possible to modify it so as to remove some of the difficulties we have mentioned.

2. Problems of the Standard Theory

There are many difficult problems associated with the hot-universe theory, but for a long time most of them seemed more metaphysical than physical, and thus of no immediate concern for scientists. The most important of these problems are:

... *The singularity problem.* The state of infinite density in which the universe was born at time 0 is called a singularity. One may wonder what there was *before* the Big Bang singularity: Where did the universe come from? If space-time itself does not exist for times less than 0, how could everything appear from nothing?

... *The flatness problem.* At school we are taught that the world is described by Euclidean geometry, and that two parallel lines never cross. When we come to the university, our professors tell us that, according to general relativity, the geometry of the universe is, in general, different from the Euclidean geometry of flat space. The universe may be closed, in the way that the surface of a sphere is closed, so that parallel lines cross one another like the meridian lines on a globe.

The only natural length parameter in general relativity is the Planck length l_P. Therefore one would expect our space to be very curved, with a typical radius of curvature on the order of 10^{-33} cm. We see, however, that our universe is just about flat on a scale of 10^{28} cm, the radius of the observable part of the universe. This 10^{10}-light-year distance to the Big Bang horizon is 60 orders of magnitude larger than l_P. Why is our universe so flat, its geometry almost exactly Euclidean?

... *The horizon problem.* This problem can be formulated in several ways. The simplest way to understand it is to consider an infinite (open or flat) universe. Why, the problem asks, did all the causally disconnected regions of the infinite universe start their expansion simultaneously (at $t = 0$)? Who gave the command?

... *The problems of homogeneity and galaxy formation.* Astronomical observations show that our universe on the very large scale is extremely homogeneous. On the scale of 10^{10} light years the distribution of matter departs from perfect homogeneity by less than a part in a thousand. Why is the universe so homogeneous? On much smaller scales, on the other hand, the universe is not homogeneous at all. It contains stars, galaxies, clusters of galaxies, voids and other structures larger than 10^8 light years. What is the origin of these important inhomogeneities?

... *The uniqueness problem.* The essence of this problem was once formulated by Albert Einstein: "What I am really interested in is whether God could have created the world differently." A few years ago it would have seemed rather meaningless to ask why our space-time is four-dimensional, why α, the fine-structure constant, is close to $\frac{1}{137}$, why the vacuum energy density (or equivalently, the cosmological constant) is so small, and so on. However, in the Kaluza-Klein and superstring theories, space-time originally has considerably more than four dimensions, but the extra dimensions have been "compactified," shrunk into thin tubes. Therefore one may wonder why compactification stopped with four effective space-time dimensions, not two or five, or some other number. Moreover, in superstring theories there are thousands of possible compactification schemes to four dimensions, each giving different low-energy particle physics (in particular, a different value of α) and a different vacuum energy density. A similar situation may occur even without recourse to compactification. In supersymmetric grand unified theories, for example, there are dozens of stable (or practically stable) vacuum states, each corresponding to a different type of symmetry breaking between fundamental interactions, and thus to a different phenomenology of low-energy particle physics. Thus it becomes rather difficult to understand why compactification and symmetry breaking have brought about just the world we know, and no other.

All these problems (and others we have not listed) are very difficult; we had no idea how to solve them within the standard hot-universe theory. That is why it is very encouraging that most of these problems, one after another, have in the last few years been either completely resolved or considerably relaxed in the context of one comparatively simple scenario of the evolution of the universe—the inflationary scenario.

3. The Inflationary Paradigm

At present there are many different versions of the inflationary-universe scenario. The main feature of all these versions, sometimes called the inflationary paradigm, is the existence of some stage of evolution at which the universe expands exponentially (or quasiexponentially) while it is in a vacuumlike state containing some (almost) homogeneous classical fields, but no (or almost no) particles. Such an expansion is what we refer to as inflation. After inflation, the initial vacuumlike state decays into particles. They interact with one another, and after the establishment of thermodynamic equilibrium the universe becomes hot. From then on it can be described by the usual hot-universe theory. (See Fig. 1.1.)

Historically, the possible existence of an inflationary stage was first envisaged by Erast Gliner (Institute for Physics and Technology, Leningrad) about 20 years ago[2]. In the 1970s I came to the realization that homogeneous classical scalar fields φ, which are present in all unified theories of elementary particles, can play the role of an unstable vacuum state, and that their decay can heat up the universe[3]. A very interesting model was proposed in 1979 by Alexey Starobinsky (Landau Institute, Moscow), who pointed out that the regime of exponential expansion and subsequent reheating of the universe occurs in the theory of gravity when one takes account of quantum corrections[4].

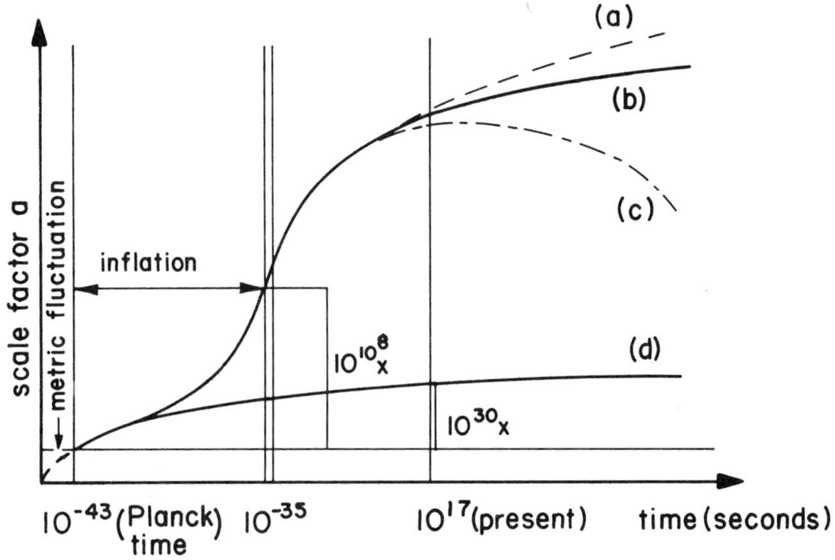

FIG. 1.1: The expansion of the universe for three possible inflationary scenarios - open (a), flat (b) and closed (c). The scale factor $a(t)$, which, at least in the case of a closed universe, can be thought of as the radius of the universe, is plotted as a function of time. The inflationary universe starts either from a singular Big Bang or from a large quantum fluctuation of a pre-existing space-time metric. After the Planck time, inflation sets in, expanding the universe by a factor of 10^{10^8} in 10^{-35} seconds. By contrast, the noninflationary hot Big Bang model (d) yields an expansion factor of only 10^{30} to the present day, 10^{17} seconds after the beginning. The inflationary epoch is followed, at 10^{-35} seconds, by a brief interlude of heating, after which the further evolution of the universe is adequately described by the standard noninflationary model.

The crucial step in the development of the inflationary paradigm was taken by Alan Guth at MIT in 1980. He suggested exploiting the stage of exponential expansion in some supercooled vacuumlike state to solve simultaneously the primordial monopole problem, the flatness problem and the horizon problem[5]. His idea was very clear and attractive, but, as Guth himself pointed out, the universe after inflation in this scenario becomes too inhomogeneous.

In October 1981 I suggested an improved version of the inflation idea, which has come to be called the new inflationary scenario[6]. This resolved some of the difficulties of Guth's original formulation and made possible the solution of some of the other cosmological problems listed here. Some months later the same idea was proposed also by Andreas Albrecht and Paul Steinhardt at the University of Pennsylvania[7], and soon the new inflationary scenario became quite popular. However, it turns out to be very difficult to realize this scenario in the context of realistic theories of elementary particles[8]. Moreover, the orthodox version of this scenario was just a modest variation on the standard hot Big Bang theory. It was still assumed that there was an initial singularity at $t = 0$, that after the Planck time t_P (or l_P/c, about 10^{-43} seconds) the universe became hot, and that inflation was just a brief interlude in the evolution of the universe.

In 1983 another inflationary scenario was proposed, which was called the chaotic inflation scenario[9]. In my opinion, this scenario is much simpler and more natural than other versions of inflation. Therefore I shall concentrate here on the chaotic inflation scenario and on some of its recent extensions, which have given rise to the idea of an eternally existing chaotic inflationary universe[10,11]. A more technical discussion of various versions of inflation can be found in a number of review articles and in some books soon to be published[8,11–14].

4. Chaotic Inflation

One of the main features of unified particle theories is that in addition to the spinor fields ψ describing electrons, neutrinos and quarks, and the vector fields A_μ mediating their interactions, these theories also contain some scalar fields φ. Scalar fields have long been used for the phenomenological description of pions. The role of scalar fields in the new theories is more fundamental and more complicated. If, for example, the potential energy $V(\varphi)$ of a scalar field φ has one minimum at some particular value φ_1, then the whole universe gradually becomes filled with the constant classical field φ_1. Such a field is almost unobservable; it looks the same to a moving observer as it does to an observer at rest, and it mimics the appearance of empty space.

However, this field can change the masses of particles. By giving different masses to vector particles mediating different interactions, such scalar fields, referred to as Higgs fields (after Peter Higgs of the University of Edinburgh), are responsible for the symmetry breaking between the weak, strong and electromagnetic interactions in grand unified theories. Another manifestation of these fields, which will be especially important for us, is connected with their potential energy density $V(\varphi)$, which in some cases may lead to the exponential (or quasiexponential, if the field φ changes slowly) expansion of the universe.

As a simplest example, let us consider the theory of a massive scalar field φ, whose quantum is a particle of nonzero mass m, much smaller than the Planck mass. Assuming that the field is minimally coupled to gravity, its potential energy density $V(\varphi)$ is simply $\frac{1}{2}m^2\varphi^2$, as illustrated in Fig. 1.2. Suppose the field φ is initially nearly homogeneous in some domain of space-time that looks locally like an expanding universe with a growing scale factor $a(t)$ and a Hubble constant $H(t)$, de-

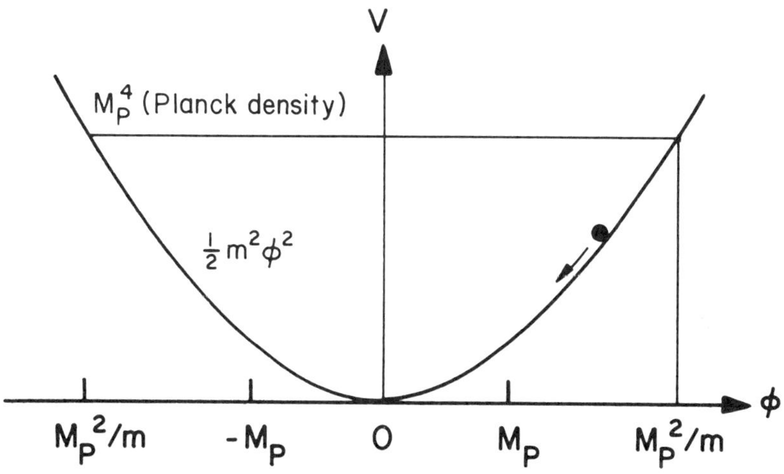

FIG. 1.2: Classical evolution of φ, the homogeneous scalar field in the chaotic inflation scenario, with a quantum of mass m, much smaller than the Planck mass M_P. For $|\varphi|$ larger than $\frac{1}{5}M_P$ (but smaller than $M_P{}^2/m$), the field (represented by the ball) rolls very slowly down toward the potential energy minimum at $\varphi = 0$. This is the inflationary epoch, during which the universe expands quasiexponentially. When $|\varphi|$ becomes smaller than $\frac{1}{5}M_P$, it oscillates rapidly about $\varphi = 0$ and its energy is transformed into heat.

fined as $\dot{a}/a$. Its evolution can then be described by the usual Klein-Gordon equation modified by cosmic expansion,

$$\ddot{\varphi} + 3H\dot{\varphi} = -m^2\varphi \tag{4.1}$$

and by the Einstein equation

$$H^2 + \frac{\kappa}{a^2} = \frac{4\pi}{3M_P^2}(\dot{\varphi}^2 + m^2\varphi^2) \tag{4.2}$$

Here κ is $+1$, -1 or 0, for a closed, open or flat universe, respectively. For the closed case ($\kappa = +1$), one can regard $a(t)$ as the literal radius of this universe. The term $3H\dot{\varphi}$ in Eq.4.1 plays the role of a friction term in the equation of motion of the field φ. Eq.4.2 tells us that if φ is initially large enough, the friction term will also be large. It can be shown[9,15] that if the initial field φ_0 is greater than $\frac{1}{5}M_P$, friction makes the variation of the field φ very slow, so that one can neglect $\ddot{\varphi}$ in equation 4.1 and $\dot{\varphi}$ in equation 4.2. In that case any solution rapidly approaches the regime

$$\varphi(t) = \varphi_0 - \frac{mM_P}{2\sqrt{3\pi}}\, t \tag{4.3}$$

$$a(t) = a_0 \exp\left(\frac{2\pi}{{M_P}^2}\left[{\varphi_0}^2 - \varphi^2(t)\right]\right) \tag{4.4}$$

From these equations it follows that for times t earlier than $\varphi_0/(mM_P)$ the universe is expanding quasiexponentially:

$$a(t) = a_0 \exp\left(H(\varphi)t\right) \tag{4.5}$$

where the Hubble "constant" is given by

$$H(\varphi) \approx \sqrt{\frac{4\pi}{3}}\,\frac{m\varphi}{M_P} \tag{4.6}$$

The exponent Ht is much larger than unity if φ exceeds the Planck mass in these early moments. This is just the inflationary regime we wanted to obtain[9].

When the field φ becomes smaller than about $\frac{1}{5}M_P$, the friction term in Eq.4.1 becomes small, and φ oscillates rapidly near its equilibrium value of zero. If this field interacts with other matter fields, these oscillations lead to abundant particle production and the heating of the universe.

The existence of an inflationary regime of this type is a general feature of a wide class of theories, including all theories in which $V(\varphi)$ varies as φ^n, higher-derivative theories of gravity, some versions of supergravity, and Kaluza-Klein theories. The role of the scalar field φ in these theories can be played by the curvature scalar, the logarithm of the radius of compactification or the like[8,14]. The remaining question to be discussed here is whether the initial conditions necessary for the realization of the inflationary regime in this scenario are sufficiently natural[15].

5. Initial Conditions

The main idea of the chaotic inflation scenario is to consider a chaotic initial distribution of the scalar field φ and to investigate its evolution without making the *ad hoc* assumption that the initial value of the field corresponds to the minimum of its potential energy $V(\varphi)$. In this sense chaotic inflation is much more general than the new inflationary universe, in which it is *assumed* that the universe *ab initio* was in thermodynamic equilibrium, and that φ was at the minimum of its temperature-dependent potential $V(\varphi, T)$.

The chaotic initial distribution of the scalar field φ is not, however, entirely unspecified. Indeed it is usually believed that for energy densities exceeding ${M_P}^4$ (or at length scales smaller than the Planck length) quantum fluctuations of the metric are so large that it becomes meaningless to speak about classical space-time; rather, one is dealing with something like a fluctuating space-time foam. Therefore when one speaks about initial conditions in some domain of a classical space-time, one automatically implies that this domain has a size greater than the 10^{-33}-cm Planck length and a density less than ${M_P}^4$. Thus $V(\varphi)$ and the squares of the spatial and temporal gradients of φ are also less than ${M_P}^4$. It follows that any initial distribution of the field φ inside any domain of a classical space-time is smooth enough that one can treat the field as essentially homogeneous within the domain.

As an example, consider first a closed universe that just appears out of the space-time foam, or from a singularity. A typical initial size of such a universe should be on the order of the Planck length ${M_P}^{-1}$, and its energy density initially should be on the order of ${M_P}^4$. Because the gradient of φ is less than ${M_P}^2$, it follows that the initial value of the field in such a universe cannot differ from some average value φ_0 by much more than M_P. One might expect that φ_0 should correspond to the minimum of $V(\varphi)$. However, fluctuations of the energy density at the moment of creation are on the order of ${M_P}^4$. Therefore the field initially "does not know" where the minimum of $V(\varphi)$ is, and later (Eq.4.3 tells us) it rolls down to this minimum very slowly. The only constraint on the initial value of φ_0 is the requirement that $V(\varphi_0)$ cannot exceed ${M_P}^4$, implying that the typical initial value φ_0 is of order ${M_P}^2/m$, much larger than any variation of φ.

Thus a typical initial distribution of the field φ inside a closed universe is essentially homogeneous, with a potential energy density of order ${M_P}^4$. If in such a universe the initial value of $(\nabla\varphi)^2$ is a quite reasonable factor of two or three smaller than $V(\varphi)$, it can be shown that the spatial gradient of φ very quickly becomes exponentially small, and the closed universe becomes an inflationary universe described by Eqs.4.1-4.6, grown from a single bubble in the space-time foam[15].

Thus we see that the initial conditions necessary for the realization of the chaotic inflation scenario in a compact (closed) universe are quite natural. In an infinite universe the situation at first glance might seem more complicated. If the initial value of φ is much larger than M_P, then, just as in the closed-universe case, φ remains rather homogeneous on a scale much larger than the Planck length, which is roughly the value of H^{-1} at that early time t. This would imply some acausal correlation between the initial values of φ in causally disconnected parts of a domain of initial size much larger than the Planck length or ct, which seems rather improbable.

The resolution of this complication is very simple[15]. There is no reason to expect that the energy density ρ can simultaneously become smaller than ${M_P}^4$ in all causally disconnected parts of a domain much larger than ${M_P}^{-1}$, since this would imply an acausal correlation. Therefore, the typical initial size of a classical space-time domain emerging from the space-time foam must be on the order of ${M_P}^{-1}$. Such domains look like separate islands of classical space-time rising out of the foam. Outside each domain the gradient of φ is not restricted to be smaller than ${M_P}^2$, and there is no acausal correlation between the values of the field in different domains. Only later can these domains become connected.

In such a case the whole universe looks like a cluster of mini-universes, some of them inflationary. Just as in the closed-universe case, a typical initial value φ inside these mini-universes would be ${M_P}^2/m$. From Eq.4.4 we see that by the time φ approaches zero the universe has expanded by a factor

$$\exp\left(2\pi {M_P}^2/m^2\right)$$

For m on the order of $10^{-7}M_P$ (which, as we shall see, is necessary for eventual galaxy formation) this implies that in our simplest model the inflationary domains of the universe typically expand to $10^{10^{14}}$ times their original size!

That much inflation is more than enough to solve many of the problems of the standard hot-universe theory. After a $10^{10^{14}}$-fold expansion, the geometry of space inside the expanding domain does not differ discernibly from the Euclidean geometry of flat space, just as the surface of a vastly expanded balloon would look very much like the surface of a flat plane. Formally, the term κ/a^2 in Eq.4.2 has become negligibly small compared with H^2 after inflation. So much for the flatness problem.

The horizon problem is also solved. The observable part of the present-day universe, with a radius of order 10^{28} cm or ten billion light years, is only a small part of the domain inflated from the smallest possible beginnings - 10^{-33} cm. Therefore we no longer have a problem about different parts of such a domain starting their expansion simultaneously. Furthermore, after expansion by a factor of $10^{10^{14}}$, all initial inhomogeneities, monopoles and domain boundaries have been swept beyond the horizon. This solves the primordial monopole problem and the homogeneity problem.

The scenario I have described is a general scheme of chaotic inflation as it can be understood at the level of classical physics. Usually one assumes that quantum effects are not essential for the description of something as large as our universe. Surprisingly, however, quantum effects play a very important role in the inflationary scenario.

6. Quantum Fluctuations

According to quantum field theory, empty space is not entirely empty. It is filled with quantum fluctuations of all types of physical fields. These fluctuations can be regarded as waves of physical fields with all possible wavelengths, moving in all possible directions. If the values of these fields, averaged over some macroscopically large time, vanish, then the space filled with these fields seems to us empty and can be called the vacuum.

In the exponentially expanding universe the vacuum structure is much more complicated[16,17]. The wavelengths of all vacuum fluctuations of the scalar field φ grow exponentially with the expanding universe. When the wavelength of any particular fluctuation becomes greater than H^{-1}, this fluctuation stops propagating, and its amplitude freezes at some nonzero value $\delta\varphi(x)$ because of the large friction term $3H\dot{\varphi}$ in the equation of motion of the field φ. (The reciprocal of the Hubble constant is ordinarily an approximation to the "age" of the universe, but during inflation it remains almost constant. In our unit system, H^{-1} also denotes the length c/H.) The amplitude of this fluctuation then remains almost unchanged for a very long time, whereas its wavelength grows exponentially. Therefore the appearance of such a frozen fluctuation is equivalent to the appearance of a classical field $\delta\varphi(x)$ that does not vanish after averaging over macroscopic intervals of space and time.

Because the vacuum contains fluctuations of all wavelengths, inflation leads to the creation of more and more new perturbations of the classical field with wavelengths greater than H^{-1}. The average amplitude of such perturbations generated during a time interval H^{-1} (in which the universe expands by a factor of e) is given by[16]

$$|\delta\varphi(x)| \approx \frac{H}{2\pi} = \frac{m\varphi}{\sqrt{3\pi}M_P} \tag{6.1}$$

Perturbations of the field lead to density perturbations that are just right for subsequent galaxy formation if m, the mass of the quantum of φ, is about $10^{-7}M_P$. The discovery of this mechanism, which gives a possible explanation of the origin of galaxies, was regarded in 1982 as one of the major successes of the inflationary scenario. Now it appears that the long-wave perturbations of φ are responsible not only for such local structure of the observable part of the universe, but also for the global structure of the universe[10,11,14].

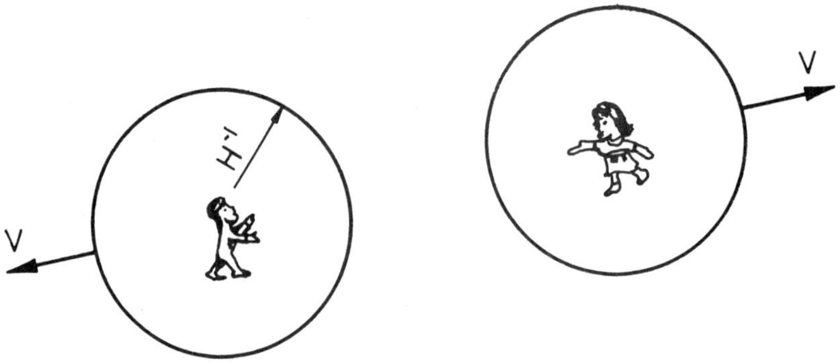

FIG. 1.3: Each observer in the inflationary universe can see only events that occur at distances closer than H^{-1}, the reciprocal of his present Hubble constant (times the speed of light, if one uses ordinary units). In that sense, every observer is surrounded by a black hole.

7. Eternal Inflation

A very unusual feature of the inflationary universe is that processes separated by distances l greater than H^{-1} proceed independently of one another. This is so because during exponential expansion any two objects separated by more than H^{-1} are moving away from each other with a velocity v exceeding the speed of light. (This does not contradict special relativity because v is not the speed of any signal; it is just the rate at which the general expansion of the universe separates two distant points.) As a result, any observer in the inflationary universe can see only those processes occurring nearer than H^{-1}, as illustrated in Fig. 1.3.

An important consequence of this general result is that the process of inflation in any spatial domain of radius H^{-1} occurs independently of any events outside it. Any two inflationary domains displaced by more than H^{-1} cannot collide or eat one another, or do each other any damage. Their expansion is due not to the annexation of the territory of their neighbors, but rather due to the peaceful (and very rapid) growth of their own volume, as allowed by general relativity. In this sense any inflationary domain of initial size exceeding $2H^{-1}$ can be considered as a separate mini-universe, expanding independently of what occurs in the rest of the cosmos.

To investigate the behavior of such a mini-universe, taking account of quantum fluctuations, let us consider an inflationary domain of initial size roughly H^{-1} containing a sufficiently homogeneous field whose initial value φ greatly exceeds M_P. Eq.4.3 tells us that during a typical time interval $\Delta t = H^{-1}$ the field inside this domain will be reduced by

$$\Delta\varphi = \frac{{M_P}^2}{4\pi\varphi} \tag{7.1}$$

If φ is much less than $\frac{1}{2}M_P\sqrt{M_P/m}$ (about $10^3 M_P$), this decrease $\Delta\varphi$ is much larger than the amplitude of the quantum fluctuations $\delta\varphi$ generated during the same time. But for larger φ (up to the classical limit of $10^7 M_P$), $\delta\varphi(x)$ will exceed $\Delta\varphi$. Because the typical wave-length of the fluctuation field $\delta\varphi(x)$ generated during this time is H^{-1}, the whole domain volume after Δt will effectively have become divided into e^3 separate domains (mini-universes) of diameter H^{-1}. In almost half of these domains the field φ *grows* by $|\delta\varphi(x)| - \Delta\varphi$, which is not very different from $|\delta\varphi(x)|$ or $H/2\pi$, rather than decreasing, as illustrated in Fig. 1.4. During the next time interval $\Delta t = H^{-1}$ the field grows again in half of these mini- universes. It can be shown that the total physical volume occupied by a permanently growing field φ increases with time like $\exp(3 - \ln 2)Ht$, and the total volume occupied by a field that does not decrease grows almost as fast as $\frac{1}{2}e^{3Ht}$.

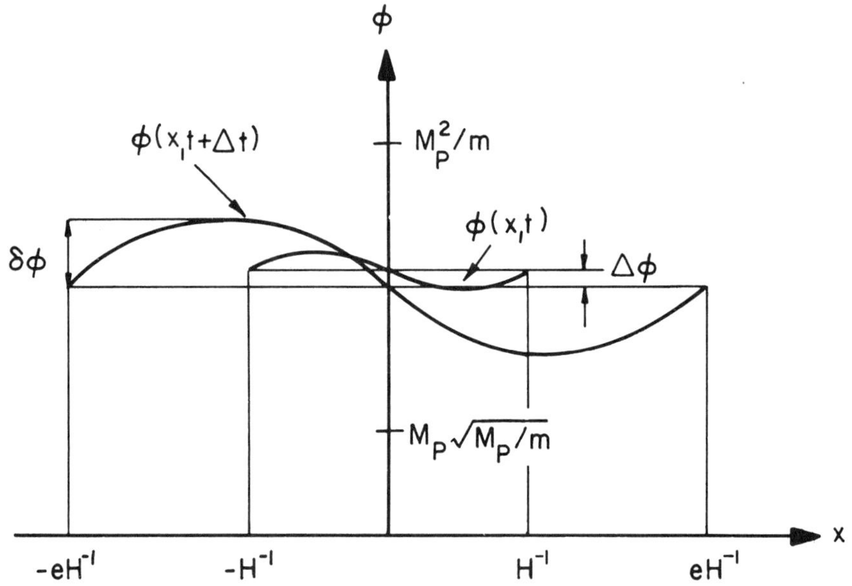

FIG. 1.4: Evolution of φ in an inflationary domain of initial radius H^{-1}. The scalar field $\varphi(x,t)$ is nearly homogeneous because quantum perturbations have wavelengths larger than the domain. In a time interval H^{-1}, the domain has an e-fold expansion. If φ is sufficiently large, its overall decrease $\nabla\varphi$ during this interval is much smaller than the fluctuating perturbation amplitudes $\delta\varphi$. Therefore the domain volume is divided into about e^3 mini-universes of radius H^{-1}, in about half of which φ will grow. The two curves show φ before and after the H^{-1} time interval.

Because the value of the Hubble constant $H(\varphi)$ is proportional to φ, the main part of the physical volume of the universe is the result of the expansion of domains with nearly the maximal possible field value, ${M_P}^2/m$, for which $V(\varphi)$ is close to ${M_P}^4$. There are also exponentially many domains with smaller values of φ. Those domains in which φ eventually becomes smaller than about $10^3 M_P$ give rise to the mini-universes of *our* type. In such domains, φ eventually rolls down to the minimum of $V(\varphi)$, and these mini- universes are subsequently describable by the usual Big Bang theory. However, the main part of the physical volume of the entire universe remains forever in the inflationary phase[10].

Similar results are also valid for the old[5,18] and the new inflationary scenarios[19], in which the main part of the volume of the universe can always remain in a state corresponding to a local extremum of $V(\varphi)$ at $\varphi = 0$. In our chaotic-inflation case[10], the results are even more surprising: Not only can the universe stay permanently on the top of a hill; it can also climb perpetually up the wall toward the largest possible values of its potential energy density, as shown in Fig. 1.5. How can it be that the universe unceasingly produces inflationary mini-universes in energetically unfavorable states with large $V(\varphi)$? What energy source supports such a process?

The answer is that the probability for a successful climb up the wall is very small indeed, but those domains in which φ jumps high enough are immediately rewarded by huge growth of their volumes. All this looks like a kind of Darwinian approach to cosmology, though one might argue that there exists an even more direct analogy between this scenario and the development of social life and economics.

The energy source that supports inflation is the gravitational energy associated with $a(t)$, the scale factor ("radius") of the universe. This gravitational energy is negative, so that the total energy of a closed universe, being the sum of the positive energy of matter and the negative gravitational energy, is zero. Just this unbounded reservoir of gravitational energy makes possible the exponentially rapid growth of the total energy of matter during inflation. The negative gravita-

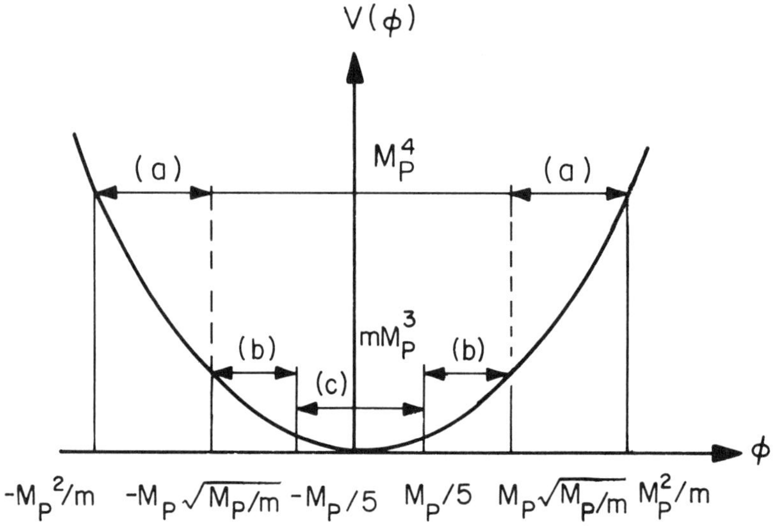

FIG. 1.5: Quantum fluctuations affect the evolution of the scalar field φ. At high field values (a) φ moves like a Brownian particle on the potential hillside $V(\varphi)$. An uphill jump is improbable, but the reward is rapid growth of the domain. Thus most of the volume of the universe comes from the expansion of domains where φ has jumped all the way up to the Planck energy density. At intermediate values (b), the upward diffusion of φ is inefficient, and its motion is described by equation (4.3) in the text. Inflation takes place while φ is in the regions (a) and (b). When φ is small (c), it oscillates near the potential minimum.

tional energy seeks any opportunity to become more negative, that is, to make inflation a nonstop process. One could say that it is the Gravitational Energy Fund that sponsors permanent production and growth of inflationary domains with large energy density.

Thus in our scenario the universe, in which there was initially at least one domain of a size on the order of H^{-1} filled with a sufficiently large and homogeneous field φ, unceasingly reproduces itself and becomes immortal. One mini-universe produces many others, and this process goes on without end, even if some of the mini-universes eventually collapse. (Everyone must die, but humanity, if it is clever enough, might exist eternally.) But this means that it is vanishingly improbable that our mini-universes would have been the first in the sequence of mini-universes illustrated in Fig. 1.6. Moreover, it no longer seems necessary to assume that there actually was some *first* mini-universe appearing from nothing or from an initial singularity at some moment $t = 0$ before which there was no space-time at all.

From general topological theorems about singularities in cosmology it does not actually follow that our universe was created *as a whole* at some moment before which the universe did not exist. The usual supposition that the whole universe appears from the unique Big Bang singularity at $t = 0$ is based on the implicit assumption that the universe *as a whole* is sufficiently homogeneous. Indeed, the observable part of our universe is very homogeneous. Observed density fluctuations are less than a part in a thousand and there has been no reason to expect that the universe is inhomogeneous on a larger scale beyond the 10^{10}-light-year horizon. In a homogeneous universe one can use the density $\rho(t)$ as a measure of time. In that case it can be shown that the universe appears *as a whole* from a singularity at $t = 0$. The initial density $\rho(0)$ is infinite, and it becomes possible to describe the whole universe in terms of classical space-time after the Planck time ${M_P}^{-1}$ (about 10^{-43} seconds), when the energy density everywhere simultaneously becomes smaller than the Planck density ${M_P}^4$.

With the invention of the inflationary scenario the situa-

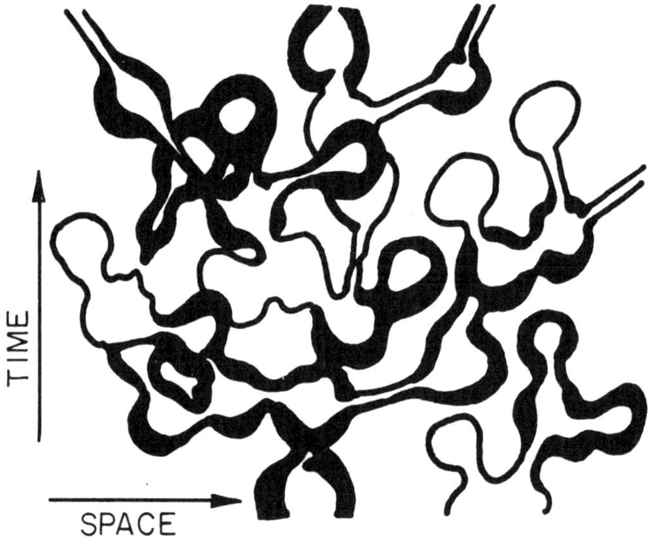

FIG. 1.6: Global structure of a chaotic, self-reproducing inflationary universe. Locally (out to the 10^{10}-light-year horizon) the universe looks quite homogeneous, but its global structure is complex. Mini- universes at the Planck energy density are "mutants" that may forget completely the "genetic code" (line thickness) of their parent universe. They may even have a different space-time dimensionality. The typical thickness of a tube connecting two mini-universes after inflation is exponentially large, but if it corresponds to a compactified inflationary universe it can be as thin as the Planck length (10^{-33} cm). If the tube then evaporates by Hawking radiation, the parent and offspring mini-universes have lost their umbilical space-time connection.

tion changes drastically. At present only inflation can explain why the observable part of the universe is so homogeneous, but from inflation it also follows that on a much larger scale the universe is extremely *in*homogeneous. In some parts of the universe the energy density ρ is now on the order of ${M_P}^4$, 125 orders of magnitude larger than the 10^{-29} or 10^{-30}g/cm^3 we can (or hope to) see. In such a senario there is no reason to assume that the universe was initially homogeneous and that all its causally disconnected parts started their expansions simultaneously.

If the universe is infinitely large (like the Friedmann open or flat universe), then it cannot have had a single beginning; a simultaneous creation of infinitely many causally disconnected regions is totally improbable. Therefore the universe cannot be infinite *ab initio*, or it must exist eternally as a huge self-reproducing entity. Some of its parts appear at different times from singularities, or may die in a singular state. New parts are constantly being created from the space-time foam when $V(\varphi)$ exceeds the Planck density, or they may revert to the foamlike state again as a result of large fluctuations in φ. But the evolution of the universe as a whole has no end, and it may have had no beginning.

8. Mutation

Darwin having been mentioned, the reader may ask: Does the universe simply reproduce itself, or are mutations also possible? To answer this question, we recall that in realistic theories of elementary particles there exist many different types of scalar fields φ_i. The potential energy $V(\varphi_i)$ often has many different local minima, in which the universe may live for an extremely long time, much greater than the 10^{10} years of our observable domain. For example, in the supersymmetric SU(5) theory, $V(\varphi_i)$ has several different minima of almost equal depth. Because the laws governing the interactions of elementary particles at the low energies at which we do experiments depend on the values of the classical fields φ_i, each of these minima corresponds to a different low-energy physics. In one of them the SU(5) symmetry between all types of interactions remains unbroken—that is, the scalar fields φ_i remain equal to zero. In other minima various symmetry breaking patterns are realized, and in only one of these minima is the broken symmetry of the weak, strong and electromagnetic interactions that which we in fact observe.

During inflation there are large-scale fluctuations of all the fields φ_i. As a result, the inflationary universe becomes divided into an exponentially large number of inflationary mini-universes, with the scalar fields taking all possible values. As inflation ends in some mini-universes, these scalar fields roll down to all possible minima of $V(\varphi_i)$. The universe becomes divided into many different exponentially large domains, realizing all possible types of symmetry breaking between the fundamental interactions. In some of these mini-universes the low-energy physics is quite different from our own. We cannot now see them because the size of our own domain is much greater than the size of its 10^{10}-light-year observable portion. We could not live in these domains because our kind of life requires our kind of low-energy physics. Therefore I would not recommend crossing the wall to a neighboring domain if it should appear over the horizon in the distant future.

It is very important that in the inflationary universe there is lots of room for all possible types of symmetry breaking and for all possible types of life. There is, therefore, no longer any need to require that in the true theory the minimum of $V(\varphi_i)$ corresponding to our type of symmetry breaking be the only one or the deepest one. This new cosmopolitan viewpoint may greatly simplify the task of building realistic models of the elementary particles.

The change of the values of the scalar fields φ_i—that is to say, the change of the vacuum state—is the simplest kind of mutation that may occur during inflation. Much more in-

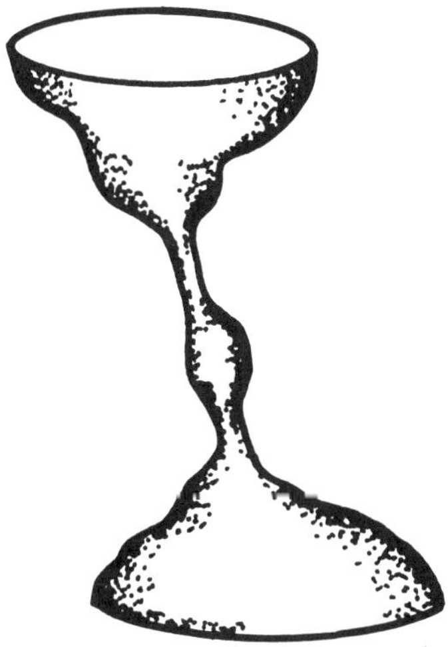

FIG. 1.7: Local compactification of decompactification of the inflationary universe alters the apparent dimensionality of space-time in a particular domain. The radius of a compactified tube can fluctuate with fluctuations of the scalar field φ.

teresting possibilities appear if one considers chaotic inflation in the higher-dimensional Kaluza-Klein theories. In those domains in which the energy density of the field φ grows to the Planck density, quantum fluctuations of the metric at a length scale of ${M_P}^{-1}$ become of order unity. In such domains an inflationary d-dimensional universe can squeeze locally into a tube of smaller dimensionality $d - n$ (or vice versa), as illustrated in Fig. 1.7. If this tube is also inflationary (in $d - n$ dimensions) and the initial length of the tube is greater than ${M_P}^{-1}$ (which is quite probable near the Planck density), then its further expansion proceeds independently of its prehistory and of the fate of its mother universe. In an eternally existing universe such processes should occur even if their probability is very small. In fact the probability of such processes is small only if $V(\varphi)$ is far below ${M_P}^4$.

Thus the inflationary universe becomes divided into different mini-universes in which all possible types of compactification produce all sorts of dimensionalities[10,11]. By this argument we find ourselves inside a four-dimensional domain with our kind of low-energy physics not because other kinds of mini-universes are impossible or improbable, but simply because our kind of life cannot exist in other domains.

This may have important implications for the building of realistic Kaluza-Klein and superstring theories[14]. For example, it is extremely complicated, if not impossible, to construct a theory in which only one type of compactification can occur, leading precisely to a four-dimensional inflationary universe with the low-energy particle physics of our experience. But from the point of view discussed here, there is no need to require that the results compactification and inflation have wrought in our realm be the only possible results, or the best. It is enough to find a theory in which such a compactification is possible. This problem is still difficult, but it is much easier than the one we have been trying to solve.

9. Cosmology and Particle Physics

One can compare cosmologists and particle physicists to two groups of people tunneling through a huge mountain from opposite sides. This analogy is not quite exact. If the two groups fail to meet they will simply make two tunnels instead of one. But if the particle physicists do not meet the cosmologists on their way through the mountain of the unknown, they will not have any complete theory at all.

The inflationary-universe scenario is a spectacular manifestation of the interplay between elementary-particle theory and cosmology. Of course inflation is not a magic word that will automatically solve all our problems and open all doors. In some theories of elementary particles it is difficult to realize inflation, whereas many other theories do not lead to a good cosmology even with the help of inflation. At the moment, however, it does seem necessary to have something like inflation to obtain a consistent cosmology at peace with particle physics.

The inflationary scenario is still developing rapidly. We are witnessing a gradual change in the overall viewpoint of cosmology. Just a few years ago there was no doubt that the universe was born in a single Big Bang singularity. It seemed absolutely clear that space-time was four-dimensional from the start, and that it remains four-dimensional everywhere. It was believed that if the universe is closed its total size is on the order of its observable part, 10^{28} cm, and that in about 10^{11} years the whole universe will collapse and disappear in a Big Crunch. If, on the other hand, the universe is flat or open, then it is infinite, and it was commonly believed that its properties everywhere are approximately the same. Such a universe would exist without end, but after the decay of its protons, predicted by the unified elementary-particle theories, there would be no baryonic matter to support life of our type. The only possible choice seemed to be between the "the hot ending" in the Big Crunch and "the cold ending" in infinite empty space.

Now it seems more likely that the universe is an eternally existing, self-producing entity, that it is divided into many mini-universes much larger than our observable portion, and that the laws of low-energy physics and even the dimensionality of space-time may be different in each of these mini-universes. This modification of our picture of the universe and of our place in it is one of the most important consequences of the inflationary scenario.

We have even gained a new understanding of why it was necessary to write a scenario when the performance is already over. The answer is that the performance is still going on, and it will continue eternally. In different parts of the universe different observers see its endless variations. We cannot see the whole play in all its greatness, but we can try to imagine its most essential parts, and perhaps ultimately understand its meaning.

Chapter 1: References

[1] Weinberg, S., *The First Three Minutes*, Basic, New York (1977).

[2] Gliner, E. B., *Zh. Eksp. Teor. Fiz.* 49, 542 (1965) [*Sov. Phys. JETP* 22, 378 (1965)]; *Dokl. Akad. Nauk SSSR*, 192, 771 (1970) [*Sov. Phys. Dokl.* 15, 559 (1970)].

[3] Linde, A. D., *Pis'ma Zh. Eksp. Teor. Fiz.* 19, 320 (1974) [*JETP Lett.* 19, 183 (1974)]; *Rep. Progr. Phys.* 42, 389 (1979).

[4] Starobinsky, A. A., *Pis'ma Zh. Eksp. Teor. Fiz.* 30, 719 (1979) [*JETP Lett.* 30, 682 (1979)].

[5] Guth, A. H., *Phys. Rev.* D23, 347 (1981).

[6] Linde, A. D., *Phys. Lett.* B108, 389 (1982).

[7] Albrecht, A., and Steinhardt, P. J., *Phys. Rev. Lett.* 48, 1220 (1982). Guth, A. H., and Steinhardt, P. J., *Sci. Am.*, May 1984, p. 90.

[8] Linde, A. D., *Rep. Prog. Phys.* 47, 925 (1984).

[9] Linde, A. D., *Phys. Lett.* 129B, 177 (1983).

[10] Linde, A. D., *Mod. Phys. Lett.* A1, 81 (1986); *Phys. Lett.* B175, 395 (1986). Goncharov, A. S., Linde, A. D., and Mukhanov, V. F., *Int. J. Mod Phys.* A2, 561 (1987).

[11] Linde, A. D., in *300 Years of Gravitation*, Hawking, S. W., Israel, W., Eds., Cambridge U. P., Cambridge, England (1987).

[12] Brandenberger, R. H., *Rev. Mod. Phys.* 57, 1 (1985). Turner, M. S., in *Proc. Cargese School on Fundamental Physics and Cosmology*, Audouze, J., and Tran Thanh Van, J., Eds., Editions Frontières, Gif-sur-Yvette, France (1986). Blau, S. K., and Guth, A. H., in *300 Years of Gravitation*, Hawking, S. W., Israel, W., Eds. Cambridge U. P., Cambridge, England (1987).

[13] Brandenberger, R. H., *Int. J. Mod. Phys.* A2, 77 (1987).

[14] Linde, A. D., *Particle Physics and Inflationary Cosmology*, Gordon and Breach, New York, to be published.

[15] Linde, A. D., *Phys. Lett.* B162, 281 (1985).

[16] Vilenkin, A., Ford, L., *Phys. Rev.* D26, 1231 (1982). Linde, A. D., *Phys. Lett.* B116, 335 (1982).

[17] Mukhanov, V. F., Chibisov, G. V., *Pis'ma Zh. Eksp. Teor. Fiz.* 10, 549 (1981) [*JETP Lett.* 33, 532 (1981)]. Hawking, S. W., *Phys. Lett.* B115, 295 (1982). Starobinsky, A. A., *Phys. Lett.* B117, 175 (1982). Guth, A. H., Pi, S.-Y., *Phys. Rev. Lett.* 49, 1110 (1982). Bardeen, J., Steinhardt, P. J., Turner, M. S., *Phys. Rev.* D28, 679 (1983).

[18] Gott, J. R., *Nature* 295, 304 (1982). Sato, K., Kodama, H., Sasaki, M., Maeda, K., *Phys. Lett.* B108, 35 (1982).

[19] Steinhardt, P. J., *The Very Early Universe*, Gibbons, G. W., Hawking, S. W., Siklos, S., Eds., Cambridge U. P., Cambridge, England (1982), p.251. Linde, A. D., *Nonsingular Regenerating Inflationary Universe*, Cambridge University preprint (1982). Vilenkin, A., *Phys. Rev.* D27, 2848 (1983).

Chapter 2

GENERATION OF DENSITY PERTURBATIONS IN INFLATIONARY UNIVERSE

1. Introduction

One of the most important problems of cosmology is the problem of the origin of the primordial density inhomogeneities which gave rise to the observable large-scale structure of the universe. Historically, three main types of such inhomogeneities were considered in the hot universe theory: adiabatic density perturbations, isothermal (or entropy) perturbations and various nonperturbatively generated inhomogeneities. Originally the corresponding terminology was related to the universe filled, at the late stages of its evolution, by ordinary (baryonic) matter and radiation[1,2].

Perturbations in baryons and radiation associated with metric perturbations (of curvature) were called adiabatic perturbations. Initial adiabatic metric perturbations h_k usually were chosen to have spectra of the type $h_k \sim k^{(n/2-2)}$ and uncorrelated phases of the partial modes. The case $n = 1$ corresponds to the flat Zeldovich (scale-free) spectrum[3]. For this spectrum the contribution of each interval from k to αk ($\alpha > 1$) to the total dispersion $4\pi \int_k^{\alpha k} \mathrm{dk}\ \mathrm{k}^2 |\mathrm{h}_\mathrm{k}^2|$ is scale-independent, and the density perturbations $\delta\rho(k)$ have the same magnitude for each k at the moment of the horizon crossing. This implies, in particular, that the density perturbations in each $\ln k$ interval are k-independent, $\delta\rho(k)/\rho = \text{const}$. As a result of gravitational instability, small initial perturbations grow. Formation of the large-scale structure of the universe is completed with a non-linear stage of this growth. A detailed description of this stage depends on the properties of the spectrum of perturbation at the beginning of this stage, $\delta\rho(k)/\rho$, which is given by

$$\frac{\delta\rho(k)}{\rho} = C(k) \left(\frac{\delta\rho}{\rho}\right)_0 , \tag{1.1}$$

where $C(k)$ is some transfer function which depends on the properties of matter in the universe. In cosmological models with massive neutrinos (or with other "hot" particles) there is a cut-off of the function $C(k)$ in the short-wave region. This gives the so-called top-down spectrum $\delta\rho(k)/\rho$, which leads to "pancake" formation. Therefore many cosmologists associate adiabatic perturbations with the "pancake" scenario of the formation of the large-scale structure of the universe[4].

Another possibility is connected with fluctuations in the composition of matter at a constant total energy density. Such perturbations were called entropy perturbations. A typical example is given by perturbations of baryon density n_B in the ultrarelativistic quark-gluon-photon plasma. Since the photon density $n_\gamma \sim T^3$ is much greater than n_B, perturbations of n_B in the very early universe do not lead to any considerable perturbations of density, metric and temperature T, and therefore such perturbations were also called isothermal or isocurvature. Being unimportant during the very early stages of the evolution of the universe, the isothermal perturbations may become very important later. Indeed, the energy density of the photon gas ρ_γ decreases as $a^{-4} \sim T^4$ during the expansion of the universe $\big(a(t)$ is the scale factor$\big)$. Meanwhile, the energy density of (nonrelativistic) protons decreases as $a^{-3} \sim T^3$. At large a the energy density of nonrelativistic matter becomes dominant, and isothermal density perturbations become important. Isothermal perturbations during the expansion of the universe give also rise to adiabatic perturbations with $C(k) \sim k^2$, see refs. 4,5. A usual assumption was that the initial spectrum of long-wave isothermal perturbation is very strongly suppressed (down-top spectrum), and therefore would lead to the hierarchical clustering scenario of galaxy formation[2].

One should note that there exist some possibilities which are intermediate between the pancake picture and the hierarchical clustering scenario. For example, in cosmological models with "cold" nonrelativistic matter and adiabatic perturbations the first objects which are formed due to gravitational instability are relatively small, but their positions are correlated with the long-wave metric perturbations[6].

It is possible also that the origin of galaxies is connected with the vortex perturbations of the metric[7] or with non-perturbative effects such as explosions of stars[8] or black hole formation[9].

To understand which inhomogeneities are actually responsible for the formation of galaxies and other large-scale objects in our universe one must investigate the possible origin of large-scale inhomogeneities in the very early universe.

Before the invention of the inflationary universe scenario[10–14] there was almost no idea which initial conditions are better, and the choice of the type of initial density perturbations was determined mainly by the physical intuition of the author. In the inflationary cosmology the situation radically changed. First of all, it was understood that all perturbations which could exist *before* inflation (if there was any stage of the universe evolution before inflation) are irrelevant for galaxy formation, since inflation washes out all initial inhomogeneities (for a review of this scenario see ref. 15). At the same time, howver, inflation leads to creation of large-scale adiabatic density perturbations with an almost flat (scale-independent) spectrum[16–24]. In some theories the amplitude of density perturbations has the magnitude $\delta\rho/\rho \sim 10^{-4}-10^{-5}$, which is just necessary for galaxy formation (whereas some other theories were rejected, since they lead to too large density perturbations after inflation).

This first success has lead to some absolutization of the results obtained. In ref. 25 a possibility for generation of isothermal perturbations was discussed in the context of the axion model, but it was claimed that such perturbations could be considerably large only in models with the amplitude of the axion field $\varphi_0 \sim 10^{15}$ GeV, whereas from the standard cosmological constraints on the field φ_0 it follows that $\varphi_0 \leq 10^{12}$ GeV. Therefore nobody has considered the possibility of the generation of isothermal perturbations during inflation quite seriously. In particular, in a recent review[26] on density perturbations generated during inflation, which was written by one of the authors of ref. 25, isothermal perturbations have not been mentioned at all. Thus, a common belief was that only adiabatic perturbations with a flat spectrum could be generated during inflation.

This conclusion is actually valid for the simplest models considered in refs. 16-24. However, this conclusion together with the statement that the energy density of the universe ρ is equal to the critical density $\rho_c(\Omega = \rho/\rho_c = 1)$, may form something like the procrustean bed for the theory of formation of the large-scale structure of the universe. Let us mention here some problems related specifically to the flat spectrum of adiabatic perturbations.

There exist some experimental constraints on the anisotropy of the cosmic microwave background radiation $\Delta T/T$. The most important constraint concerns the large-scale (quadrupole) anisotropy, $(\Delta T/T)_Q \leq 2 \times 10^{-5}$, see ref. 27. Theoretical estimates of $(\Delta T/T)_Q$ at present are considerably author-dependent[28–31],* but many authors agree that the theoretical predictions of $(\Delta T/T)_Q$ for the neutrino-dominated universe are at the verge of contradiction with the observational limit $(\Delta T/T)_Q \leq 2 \times 10^{-5}$. Another problem of the neutrino-dominated models is the large value of the dispersion of velocities of galaxies, $\sigma_v \sim 10^3$ km/s, which exceeds considerably the observed value $\sigma_v \sim$ 300-400 km/s. The main difficulty of the standard neutrino scenario is a too rapid evolution of the large-scale structure during the nonlinear stage[32–34], which was a reason for the development of a new generation of cosmological models based on theories with unstable neutrinos[35,36]. However, in the models with unstable neutrinos the value of $(\Delta T/T)_Q$ is approximately the same as in the standard massive neutrino scenario[37] – this value may be too large.

In the cold dark matter the dispersion of velocities is also too large, $\sigma_v \sim$ 800 km/s [38]. This difficulty may be eliminated by the introduction of a small cosmological constant Λ, which somewhat improves the cold dark matter model[39]. Still the possibility of obtaining the observed large-scale structure of the universe in a universe with cold dark matter and with $\Omega = 1$ is far from being clear. A more extensive list of difficulties associated with formation of the large-scale structure of the universe from adiabatic perturbations with a scale-free

---

* However, recently some agreement has been attained (see ref. 92).

spectrum can be found in any of the recent reviews of this problem, see, *e.g.*, ref. 40.

These difficulties have stimulated a search for an alternative source of initial density perturbations. A rather interesting possibility suggested in the early 80's is based on the theory of strings[41,42,43], which are formed[44] during high-temperature phase transitions in the very early universe[45]. According to recent estimates[46], strings with mass per unit length $\mu \sim (2 \times 10^{16} \text{GeV})^2$ give rise to density perturbations sufficient for galaxy formation and to relatively small anisotropy of the cosmic microwave background radiation, $\Delta T/T \sim 2 \times 10^{-5}$. It remained unnoticed, however, that heavy strings with $\mu \sim (2 \times 10^{16} \text{GeV})^2$ could be produced during the high-temperature phase transitions only if the reheating temperature after inflation, T_R, is extremely high. Typically it requires $T_R \sim 0(\mu) \sim 10^{16}$ GeV[45]. However, in most of the models of inflation T_R is much smaller than 10^{16} GeV[15], (though some exceptions to this rule do exist).

It may seem therefore that in order to buy heavy strings formed during high-temperature phase transitions[41–44] one must first sell inflation. Such a price at present seems too high, and it would be desirable to find another way to resolve potential difficulties associated with adiabatic perturbations with a flat spectrum. Of course, one may argue that one should not consider these difficulties very seriously because of the large uncertainty both in experimental data and in theoretical estimates based on the theory of galaxy formation. Nevertheless, it would be unwise to forget about these problems and not to study their possible solutions, since they may shed new light on the problem of formation of the large-scale structure of the universe. Moreover, in our opinion, it would be naive to expect that the great variety of large-scale objects in the universe (galaxies, clusters, superclusters, quasars, etc.) have originated from a single type of perturbation, either from adiabatic ones with a flat spectrum or from perturbations generated by strings.

In what follows we will discuss another class of inhomogeneities which may play an important role in the inflationary cosmology. Namely, in refs. 47-50 it was shown that not only adiabatic, but also isothermal perturbations can be generated during inflation, and in a wide class of elementary particle theories these perturbations become dominant during the later stages of the evolution of the universe. This fact by itself would not be very important (see below); what is important, however, is the possibility to obtain isothermal perturbations with a spectrum, which may differ considerably from the flat one[47,48,50]. This may lead to many interesting consequences. It is possible also to obtain a spectrum of adiabatic perturbations which decreases in the long-wave region, and which may lead to a considerable suppression of the quadrupole anisotropy $(\Delta T/T)_Q$. It is possible to obtain a spectrum which has several different maxima, etc. Moreover, in some theories, nonperturbative inhomogeneities such as exponentially large bubbles, strings, domain walls etc. are also generated during inflation. This opens up new possibilities in the theory of the large-scale structure of the universe.

To explain the origin of isothermal inhomogeneities generated during inflation, let us first recall the main idea of the theory of generation of density perturbations in the inflationary cosmology. The simplest models of inflation considered in refs. 16-24 describe the evolution of a single scalar field Φ (the so-called inflaton field) which drives inflation. The main part of the energy density during inflation is the potential energy density $V(\Phi)$ of this field. Inflation gives rise to long-wave perturbations of this field, which later transform to perturbations of density (temperature) of particles created during the decay of the classical inflaton field Φ. Note that perturbations generated in the models considered in refs. 16-24 were purely adiabatic. For example, the ratio n_B/n_γ was determined only by the parameters of the theory responsible for the baryogenesis, *i.e.* no perturbations of n_B/n_γ were generated after inflation.

However, in realistic elementary particle theories there exist scalar fields φ of many different types. During inflation perturbations of all these fields were generated. The average energy density of these fields initially was smaller than the energy of the inflaton field Φ (otherwise one of these fields would drive inflation and would be called "inflaton"). However, in some cases the density *perturbations* of these fields may be greater than density perturbations of the inflaton field Φ. Moreover, if some of the fields φ interact sufficiently weakly with other fields, their energy density in the expanding universe decreases more slowly than the energy density of the hot dark matter created during the decay of the inflaton field Φ. As a result, the density inhomogeneities connected with perturbations of the fields φ may become dominant. Such perturbations appear even if there were no perturbations of the inflaton field Φ and no temperature perturbations (which are proportional to the perturbations of the inflaton field). Therefore the inhomogeneities related to the fields φ can be called "isothermal"[47–49] or "isoinflaton"[50].

We wish to emphasize that the existence of the fields φ, which very weakly interact with other fields ("hidden sector" of the theory), is a typical feature of many realistic theories of elementary particles. One such field is the well known axion field, which is necessary for a natural solution of the strong *CP* problem[51]. Another example is the so-called Polonyi field, which is a necessary ingredient of all models of elementary particles based on $N = 1$ supergravity[52]. The fields of both these types are present also[53] in superstring theories[54,55], which are so popular now. In many such theories these "hidden" fields rather than the particles produced by the decay of the inflaton field Φ give the dominant contribution to the present energy density of the universe. (What could be more natural than to expect that the hidden mass of the universe is connected with hidden fields[56,57].) This explains the possible importance of isothermal perturbations during the later stages of the evolution of the universe.

One should note that from the point of view of the theory of large-scale structure formation, the difference between adiabatic and isothermal density perturbations with a flat spectrum is not very large. Such isothermal perturbations finally give rise to adiabatic perturbations with a nearly flat spectrum, which may lead again to the "pancake" scenario. The magnitude of the quadrupole anisotropy of $\Delta T/T$ in such a case would be even 5 times greater than in the usual adiabatic theory[58,50,59]. Therefore the potential importance of the isothermal perturbations stems not from their "isothermal" origin, but from the possibility of obtaining different spectra of perturbations by a proper choice of the underlying theory of elementary particles. This makes possible a more flexible approach to the theory of formation of large-scale structure of the universe.

Some other effects discussed in this chapter are connected with phase transitions which may occur during the last stages of inflaton (*i.e.* out of thermal equilibrium). These phase transitions may occur due to the change of the inflaton field Φ and the Hubble parameter H during inflation. Such changes are especially large in the chaotic inflation scenario[14], which (together with its different versions, connected with the gravitational vacuum polarization[60], $N = 1$ supergravity[61], quantum creation of the universe with $V(\varphi) \sim M_P^4$ [62] and Kaluza-Klein inflation discussed in ref. 63) seems to be the only viable scenario of inflation which exists at present. (For a discussion of the problems associated with the new inflationary universe scenario[13] see refs. 15,64 and Sec. 2 of this chapter.) According to the simplest version of the chaotic inflation scenario[14], the inflaton field Φ changes by several units of $M_P \sim 10^{19}$ GeV during the last stages of inflation, during which density perturbations responsible for the structure of the observable part of the universe are generated. Such a large change of the field Φ may lead to various phase transitions, similar to the high-temperature phase transitions[45] mentioned above. This may lead to creations of strings, bubbles and other objects without any need for high-temperature phase transitions. These objects become exponentially large (but smaller than the observable part of the universe) during the last stages of inflation. This effect, especially the effect of creation of exponentially large bubbles containing matter in different phases, may play a very important role in the theory of formation of the large-scale structure of the universe. As a byproduct, such phase transitions may lead to a reasonably small monopole density after inflation.

This chapter is organized as follows. In Sec. 2 we recall some features of the inflationary universe scenario and make some numerical estimates which will be useful for the understanding of the subsequent materials. In Sec. 3 we discuss the theory of generation of the long-wave perturbations of a scalar field φ and of its energy density during inflation. In Sec. 4 we consider a simplest model of the inflationary universe containing two noninteracting scalar fields Φ and φ[47,48]. This example will make it possible to explain the relation between adiabatic and isothermal perturbations generated during inflation. In Sec. 5 we discuss isothermal perturbations in the theory of axions[47–50], in which there exists a rather nontrivial mechanism of suppression of the long-wave density perturbations[47,48,50]. This mechanism is connected with a periodicity of the effective potential $V(\varphi)$ of the axion field φ. In Sec. 6 we discuss the effects connected with phase transitions which occur due to a slow decrease in the inflaton field $\Phi(t)$ at the last stages of inflation. These effects may also lead to the appearance of several different maxima in the spectrum of adiabatic perturbations. In Sec. 7 we consider nonperturbative effects which lead to the formation of exponentially large strings, domain walls and bubbles during inflation. The main results of our investigation are discussed in the conclusions.

2. Inflation

Let us consider a flat Friedmann universe with metric

$$ds^2 = dt^2 - a^2(t)(dx^2 + dy^2 + dz^2) \tag{2.1}$$

which satisfies the Einstein equation

$$H^2 \equiv \left(\frac{\dot{a}}{a}\right)^2 = \frac{8\pi}{3M_P^2}\,\rho\ . \tag{2.2}$$

Here $M_P \sim 10^{19}$ GeV is the Planck mass and ρ is the energy density of matter. If during some period of time ρ changes very slowly, the Hubble parameter H also changes slowly ($\dot{H} \ll H^2$), and the universe expands quasiexponentially,

$$a(t) = a_0 \exp\left(\int_{t_0}^{t} dt H(t)\right)\ . \tag{2.3}$$

Such a regime of expansion is called inflation. In particular, if $H = \text{const}$,

$$a(t) = a_0 e^{H(t-t_0)}\ . \tag{2.4}$$

Originally, the most popular version of this scenario was the new inflationary universe scenario[13] based on the theory of high temperature phase transitions from the state $\Phi = 0$ with energy density $\rho = V(\Phi = 0)$, where $V(\Phi)$ is the effective potential of the classical scalar field Φ. However, there exist at least two large difficulties which preclude a successful realization of this scenario.

(i) Density perturbations generated during inflation are small enough ($\delta\rho/\rho \sim 10^{-5}$ only if the inflaton field Φ is a very weakly interacting field ($\lambda \sim 10^{-14}$ for the theory $\frac{1}{4}\lambda\Phi^4$). High-temperature corrections for the effective potential $V(\Phi, T)$ of such a field are typically extremely small, and they cannot lead to any phase transition during a small time interval $0(H^{-1}_{(T)})$ prior to inflation[15,64].

(ii) A typical closed universe has a lifetime $\tau \sim 0(M_P^{-1})$. Therefore it collapses before inflation starts at $t \sim H^{-1}$ *if inflaton starts at the energy density* $\rho \ll M_P^4$. In all version of the new inflationary universe scenario considered so far $\rho \ll 10^{-10} \times M_P^4$. Higher values of $\rho = V(0)$ typically lead to density perturbations $\delta\rho/\rho \gg 10^{-4}$ after inflation. Therefore it does not seem possible to obtain a solution of the flatness problem in this scenario, at least for a closed universe[16]. Despite many efforts, these problems of the new inflationary universe scenario still remain unresolved. Fortunately, the corresponding problems disappear in the chaotic inflation scenario[14]. This scenario is based on the investigation of the evolution of a nonequilibrium distribution of the inflaton field Φ. As is argued in refs. 14,64, a typical distribution of the field Φ in a *classical* universe, which emerges from the space-time foam at the Planck time, satisfies the conditions

$$\begin{aligned} (\partial_0\Phi)^2 &\sim M_P^4 \ , \\ (\partial_i\Phi)^2 &\sim M_P^4 \ , \\ V(\Phi) &\sim M_P^4 \ . \end{aligned} \tag{2.5}$$

A typical initial size of a closed universe is $0(M_P^{-1})$, which is equal to the size of the horizon at that time. Since $(\partial_0\Phi)^2$, $(\partial_i\Phi)^2$

and $V(\Phi)$ initially are of the same order $0(M_P^4)$, there is a sufficiently large probability that initially $(\partial_0\Phi)^2$, $(\partial_i\Phi)^2 < V(\Phi)$ inside a closed universe of the size $0(M_P^{-1})$ (which is the smallest size that can be discussed in terms of classical space-time). For a wide class of potentials $V(\Phi)$ including, *e.g.*, any potential $V \sim c\Phi^n$, it can be shown that in such a universe the field Φ soon becomes almost exactly homogeneous, $(\partial_i\Phi)^2 \ll V(\Phi)$, and $(\partial_0\Phi)^2$ becomes much smaller than $V(\Phi)$ for $\Phi \geq M_P$. Consequently, the value of $V(\Phi)$ for $\Phi \geq M_P$ changes in time very slowly, and the universe expands quasi-exponentially[14]. Therefore, for the realization of the chaotic inflation scenario, there is no need to satisfy numerous unnatural constraints, which are imposed on the shape of the effective potential in the new inflationary universe scenario. Since a typical closed universe at the Planck time consists of just *one* causally connected region of a size $0(M_P^{-1})$, no problems with causality mentioned in [65] arise in the chaotic inflation scenario. Moreover, the initial condition $(\partial_0\Phi)^2 \leq V(\Phi) \sim M_P^4$ actually can be relaxed, since the value of $(\partial_0\Phi)^2$ decreases in an expanding universe much more rapidly than $V(\Phi) \sim \Phi^n$, and soon $(\partial_0\Phi)^2$ becomes much smaller than $V(\Phi)$ even if initially $(\partial_0\Phi)^2 \gg V(\Phi)$ [60,64,66,67]. Thus, the chaotic inflation scnenario can be realized in a wide class of theories under rather natural initial conditions at the very early stages of the universe evolution.

For a more detailed discussion of the chaotic inflation scenario and of the initial conditions sufficient for its realization, see refs. 14,15,60,64,66,67. We would like to emphasize that the inflaton field Φ in the chaotic inflation scenario does not have to be an elementary scalar field. The curvature scalar R in a theory with terms $\sim R^2$ [11,60], the compactification radius in Kaluza-Klein theories[63], or a condensate of fermions $\langle\bar{\psi}\psi\rangle$ can play the same role as the field Φ. Most of the results obtained in the present paper do not depend on a particular realization of the inflationary universe scenario. However, since the chaotic inflation scenario at present seems to be the best scenario on the market, we will list below some useful relations for the simplest versions of this scenario based on theories with potential $V(\Phi) = c\Phi^n$ for a field Φ minimally coupled to gravity.

After a small initial period of inflation the field Φ becomes almost homogeneous and its evolution is described by the equation

$$\ddot{\Phi} + 3H\dot{\Phi} = -\frac{dV}{d\Phi} = cn\Phi^{n-1} , \tag{2.6}$$

where

$$H^2 = \left(\frac{\dot{a}}{a}\right)^2 = \frac{8\pi}{3M_P^2}\left(\frac{1}{2}\dot{\Phi}^2 + c\Phi^n\right) . \tag{2.7}$$

During inflation the field Φ decreases in time. For example, in the theory $V(\Phi) = \frac{1}{4}\lambda\Phi^4$

$$\Phi(t) = \Phi_0 \exp\left(-\sqrt{\frac{\lambda}{6\pi}}\, M_P t\right) . \tag{2.8}$$

From Eq.2.8 it follows that $\frac{1}{2}(\dot{\Phi})^2 \le \frac{1}{4}\lambda\Phi^4$ for $\Phi \ge \frac{1}{3}M_P$, which implies that inflation occurs $\Phi \ge \frac{1}{3}M_P$. The scale of the universe during inflation grows as follows:

$$a(t) = a_0 \exp\left(\frac{4\pi}{n}\left(\Phi_0^2 - \Phi^2(t)\right)\right) . \tag{2.9}$$

It is worth noting that Eq.2.9 is almost model-independent. It does not depend on c, and it has almost the same functional form for different n.

Let us consider again the theory $V(\Phi) = \frac{1}{4}\lambda\Phi^4$. As was shown in refs. 21-24,60,15, adiabatic density perturbations are given by $\delta\rho/\rho \sim 10^{-5}$ for $\lambda \sim 10^{-14}$. This means that at the end of inflation ($\Phi \sim \frac{1}{3}M_P$)

$$(2.10) \qquad H \sim \sqrt{\frac{2\pi\lambda\Phi^4}{3M_P^2}} \sim 2 \cdot 10^{-8} M_P \sim 2 \cdot 10^{11} \text{ GeV} .$$

If adiabatic perturbations are smaller than $\delta\rho/\rho \sim 10^{-5}$, the corresponding value of H at the end of inflation should be smaller than 2×10^{11} GeV. In what follows we will assume that this is indeed the case. The energy density at the end of inflation

$$(2.11) \qquad \rho \sim 2V\left(\frac{1}{3}M_P\right) \sim \frac{1}{2}\lambda\left(\frac{1}{3}M_P\right)^4 \sim 10^{-16} M_P^4 .$$

If this energy would rapidly transform into heat, the reheating temperature T_R would be rather large.

$$(2.12) \qquad \max(T_R) \approx 10^{-5} M_P \sim 10^{14} \text{ GeV} .$$

However, the reheating temperature typically is some orders smaller than 10^{14} GeV because of the very slow decay of the weakly interacting classical field Φ into usual particles. As we have mentioned already, this solves the primordial monopole problem[68], but makes it very difficult to produce heavy strings after inflation.

The universe after inflation expands approximately $0(T_R/T_0)$ times, where $T_0 \sim 3^o K$. This gives $0(T_R/T_0) \sim 10^{28}$. (An error by two or three orders of magnitude is possible, since reheating does not occur immediately after inflation.) Therefore the size of that part of the universe which is observable now (the size of the horizon, $R_H \sim 10^{28}$ cm) at the end of inflation was given by $\tilde{R}_H \sim 1$ cm. The value of H^{-1} at that time was $0(10^{-25}$

cm). Perturbations of the length $\tilde{R}_H$ were generated at the time t, at which the field Φ was given by the equation

$$\tilde{R}_H \sim H^{-1} \exp\left(\frac{4\pi\Phi_H^2}{nM_P^2}\right) \tag{2.13}$$

(see Eq.2.9 and the next section). This yields

$$\Phi_H \sim \sqrt{5n}\, M_P \, . \tag{2.14}$$

For the theory $\frac{1}{4}\lambda\Phi^4$ this gives

$$\Phi_H \sim 4.5 M_P \sim 5 \times 10^{19}\text{GeV} \, . \tag{2.15}$$

Another interesting scale is the galaxy scale $l_g \sim 10^{22}$ cm, or $\tilde{l}_g \sim 10^{-6}$ cm. From Eq.2.9 it follows that such perturbations were generated in the theory $\frac{1}{4}\lambda\Phi^4$ at

$$\Phi_g \sim 4M_P \, . \tag{2.16}$$

This means that all effects responsible for galaxy formation occur at $4M_P < \Phi_g(t) \leq 4.5M_P$.

The Hubble parameter $H(\Phi_0)$ is given by

$$H(\Phi) \sim \sqrt{\frac{2\pi\lambda\Phi_g^4}{3M_P^2}} \sim 3 \times 10^{-6} M_P \sim 3 \times 10^{13}\text{GeV} \, . \tag{2.17}$$

During the decrease of the field Φ from Φ_H to Φ_g the Hubble parameter does not change very much, $H(\Phi_g) \sim 0.8H(\Phi_H)$. Therefore in many cases the value of H in the theory of generation of density perturbations can be considered approximately constant. However, $H^2(\Phi_g) \sim 0.6H^2(\Phi_H)$. This means that one should be careful in cases, in which effects appear which depend on the value of $H^2(\Phi)$.

The estimates and expressions contained in this section are model-dependent. They may give some idea of a possible behaviour of $\Phi(t)$ and $H(\Phi)$. However, one should bear in mind that in different models one can obtain different numerical values of Φ and $H(\Phi)$ at the end of inflation.

3. Scalar Field Fluctuations in the Inflationary Universe

Generation of density perturbations in the inflationary universe[16–24] is deeply connected with the generation of long-wave quantum fluctuations of the scalar fields φ during inflation[69,70,18]. To understand the physical meaning of this effect let us first recall the well-known equation for fluctuations $\langle(\delta\varphi)^2\rangle$ of the field φ in Minkowski space:

$$\langle(\delta\varphi)^2\rangle = \frac{1}{(2\pi)^3}\int_0^\infty \frac{d^3k}{\sqrt{k^2+m^2}}\left(\frac{1}{2}+\left\langle a_k^\dagger a_k\right\rangle\right) , \tag{3.1}$$

where the first term corresponds to vacuum fluctuations, and $n_k = \langle a_k^\dagger a_k\rangle$ is the density of φ-particles. For example, in a thermodynamic equilibrium state

$$n_k = \left[\exp\left(\frac{\sqrt{k^2+m^2}}{T}\right)-1\right]^{-1} . \tag{3.2}$$

Now let us consider the inflationary universe with a scale factor $a(t) = e^{H(t-t_0)}$ containing a scalar field φ with an effective potential $V(\varphi)$ with a sufficiently small mass $m^2 = \mathrm{d}^2V/\mathrm{d}\varphi^2 \ll H^2$. As it was shown in refs. 69,70,18,71, in such a case after a sufficiently long period of inflation (see below) the value of

$\langle(\delta\varphi)^2\rangle$ can be approximately represented as a sum

(3.3)

$$\langle(\delta\varphi)^2\rangle \simeq \frac{1}{(2\pi)^3}\int\limits_0^\infty \frac{d^3k}{2\sqrt{k^2+m^2}} + \frac{H^2}{4\pi^2}\int\limits_0^H \left(\frac{k}{H}\right)^{2m^2/3H^2} \mathrm{d}\ln k \ .$$

An interpretation of the last term in (3.3) is very interesting. A comparison of (3.1), (3.2) and (3.3) suggests that in an exponentially expanding universe

$$n_k = \frac{H^2\sqrt{k^2+m^2}}{2k^3}\left(\frac{k}{H}\right)^{2m^2/3H^2} . \tag{3.4}$$

Thus, $n_k \sim H^2 m/k^3 \gg 1$ for $k \ll m, H$. Note, that for $n_k \gg 1$ the anticommutativity of creation and annihilation operators $a_k^\dagger$ and a_k becomes unimportant ($[a_k a_k^\dagger] = 1, \langle a_k^\dagger a_k\rangle = n_k \gg 1$), and the long-wave quantum fluctuations of the field φ can be considered as being *classical* fluctuations $\delta\varphi(k)$ of the field φ. This effect is very similar to the effect of Bose condensation, where n_k becomes infinite for $k = 0$.

The differences and similarities between these two effects become manifest if one considers a spatial distribution of the field φ in the inflationary universe. As it is noted in the previous section, the classical part φ_c of the field φ after inflation becomes homogeneous, $\varphi_c = \varphi_c(t)$, and satisfies the equation

$$\ddot{\varphi}_c + 3H\dot{\varphi}_c = -\frac{\mathrm{d}V(\varphi)}{\mathrm{d}\varphi} \tag{3.5}$$

whereas small perturbations $\delta\varphi$ of this field satisfy the equation

$$\delta\ddot{\varphi} + 3H\delta\dot{\varphi} - e^{-2Ht}\Delta(\delta\varphi) = -\frac{\mathrm{d}^2V}{\mathrm{d}\varphi_c^2}\delta\varphi \equiv -m^2(\varphi_c)\delta\varphi \ . \tag{3.6}$$

This equation implies that the wavelength of a small perturbation $\delta\varphi$ grows in time as $\exp(Ht)$ (k decreases as $\exp(-Ht)$).

Perturbations with $k \gg H \gg m^2$ oscillate as $\sin kt$, whereas perturbations with $k \ll H$, $m^2 \ll H^2$ do not oscillate and their magnitude remain almost frozen during inflation due to the large friction term $3H\delta\dot{\varphi}$ in (3.6). Thus, the amplitude of the field φ at some point x is given by a sum of the classical field $\varphi_c(t)$, of the "frozen" long-wave semi-classical perturbations with $k < H$ and of (oscillating) quantum fluctuations with $k > H$. The last contribution (almost) completely disappears after the renormalization of $\langle(\delta\varphi)^2\rangle$. This corresponds to the subtraction of the first term from Eq.3.3.:

$$(3.7)\quad \sigma^2 \equiv \langle(\delta\varphi)^2\rangle \simeq \frac{H^2}{4\pi^2}\int_0^H \left(\frac{k}{H}\right)^{2m^2/3H^2} d\ln k = \frac{3H^2}{8\pi^2 m^2}\ .$$

The contribution to σ^2 from the fluctuations with momenta $k_0 < k \ll H$ is given by

$$\sigma^2_{k_0} = \langle(\delta\varphi)^2\rangle_{k_0} = \frac{H^2}{4\pi^2}\int_{k_0}^H \left(\frac{k}{H}\right)^{2m^2/3H^2}$$

$$d\ln k = \frac{3H^4}{8\pi^2 m^2}\left(1 - \left(\frac{k_0}{H}\right)^{2m^2/3H^2}\right)\ .$$

(3.8)

This means that the mean contribution to $\langle(\delta\varphi)^2\rangle$ from the fluctuations in a logarithmic interval $\Delta\ln k = 1$ is given by

$$(3.9)\qquad \delta\varphi(k) = \frac{H}{2\pi}\left(\frac{k}{H}\right)^{m^2/3H^2}\ .$$

An important feature of $\delta\varphi(k)$ is that $\delta\varphi(k)$ is scale-independent for $m^2 \ll H^2$, $\delta\varphi(k) \sim H/2\pi = \text{const}$.

Now we are in a position to understand the spatial distribution of the field φ (neglecting the short-wave quantum fluctuations, as discussed above). Indeed, the dispersion $\sigma = \sqrt{\frac{3}{2}} H^2/2\pi m$ (3.7) gives the average deviation of the field φ from its classical value φ_c. By comparing σ^2 and $\langle(\delta\varphi)^2\rangle_{k_0}$ one can deduce that the main contribution to $\langle(\delta\varphi)^2\rangle$ is given by fluctuations with exponentially large wave-lengths, $\tilde{l} \sim \tilde{k}^{-1} \sim H^{-1}\exp(3H^2/m^2)$. This means that on scales smaller than $\tilde{l}$ the field φ changes very slowly. It looks like an almost homogeneous classical scalar field with an average amplitude

$$\tilde{\varphi} = \sqrt{\varphi_c^2 + \langle(\delta\varphi)^2\rangle} \ . \tag{3.10}$$

However on scales $l \gg \tilde{l}$ the magnitude of the field φ changes considerably and, for $\varphi_c = 0$, it may vary between $-0(\tilde{\varphi})$ and $+0(\tilde{\varphi}) = 0(\sigma)$.

It may be instructive to consider the particular example of a massless field φ. In this case the integral (3.7) formally diverges. However this divergence disappears if one takes into account that inflation starts at some time t_0. In this case it can be shown that the integration actually should start not with $k = 0$, but with some $k_{\min} \sim l_{\max}^{-1} \sim H\exp(-H(t - t_0))$ which yields[69,70,18]

$$\sigma^2 = \frac{H^3}{4\pi^2}(t - t_0) = \frac{H^2}{4\pi^2}\ln(l_{\max}H) \ . \tag{3.11}$$

Here $l_{\max}$ corresponds to the long-wave cut-off of the spectrum of perturbations generated during inflation, which starts at $t = t_0$. In most versions of the inflationary universe scenario $l_{\max}$ is much greater than the size of the observable part of the universe $l_u \sim 10^{28}$ cm.

This result has the following interpretation. During the time interval τ all quantum fluctuations of the field φ with momenta $He^{-H\tau} \leq k \leq H$ become "stretched" by the expansion of the universe , so that their momenta become smaller than H^{-1}. Perturbations with $k < H$ also become stretched, but their amplitudes remain frozen. Since perturbations with different wavelengths give independent contributions to $\sigma^2 = \langle(\delta\varphi)^2\rangle$, the "stretching" of perturbations with $k > H$ serves as a source of linear growth of σ^2 in time (3.11).

One may consider this process as Brownian motion of the field φ. Roughly speaking, for each time interval $\Delta t \simeq H^{-1}$ the logarithm of wavelengths of all perturbations grows by $\Delta \ln k = 1$. The previous distribution of the classical long-wave field φ stretches, but it also receives a new contribution $\delta\varphi \sim H \sin \overline{kx}/\sqrt{2\pi}$ with $k \sim H$. Integrated over x this just yields the following contribution to σ^2

$$\Delta\left(\langle(\delta\varphi)^2\rangle\right) \sim (\delta\varphi)^2 \sim \frac{H^2}{4\pi^2} \tag{3.12}$$

during the time interval $\Delta t \sim H^{-1}$, in agreement with (3.11).

According to (3.11) the average amplitude of the massless field φ grows in time as $\sigma \sim H\sqrt{H(t-t_0)}/2\pi$. The main contribution to $\langle(\delta\varphi)^2\rangle$ is given by perturbations with an exceptionally large wavelength $l \sim H^{-1}\exp\left(H(t-t_0)\right) \sim l_{\mathrm{max}}$.

Perturbations of the field φ give rise to the density perturbations $\delta\rho$. If the field φ changes in time sufficiently slowly, its energy density ρ_φ is given by $V(\varphi)$. A useful relation which is valid for any $V(\varphi) \sim \varphi^n$ is

$$\frac{\delta\rho_\varphi(k)}{\rho_\varphi} \sim \frac{\delta V\left(\varphi(k)\right)}{V(\varphi)} \sim \frac{\delta\varphi(k)}{\varphi} \sim \frac{H}{2\pi\varphi}\left(\frac{k}{H}\right)^{m^2/3H^2} \tag{3.13}$$

Here the value of the field φ includes φ_c together with the contribution from all fluctuations with momenta smaller than k. Since the sum of such perturbations changes at distances greater than k^{-1}, the value of φ in (3.13) also will depend on x if the dispersion of long-wave perturbations $\tilde{\sigma}_k \equiv \sqrt{\sigma^2 - \sigma_k^2}$ is comparable with φ_c. Eq. (3.13) is valid only if φ does not change considerably on a distance $\Delta l \sim k^{-1}$. This occurs if φ is larger than the short-wave dispersion σ_k (3.8). In Sec. 5 we will consider the axion model for which this condition is not satisfied.

Eq. (3.13) implies that for $m^2 \ll H^2$ the spectrum $\delta\rho_\varphi(k)/\rho_\varphi$ is almost exactly scale independent.* If the field φ is actually the inflaton field Φ (the field which drives inflation) then the total energy density of matter is $\rho_{\text{tot}} \simeq \rho_\Phi \simeq V(\Phi)$. The inhomogeneities of $V(\Phi)$ after the deca y of the inflaton field Φ give rise to adiabatic perturbations of $\rho_{\text{tot}} \sim T^4$, $\delta\rho_{\text{tot}}/\rho_{\text{tot}} \sim \delta\rho_\Phi/\rho_\Phi \sim \delta T/T$. For an estimate of the magnitude of $\delta\rho_\Phi/\rho_\Phi$ one can use Eq. (3.13). However, one should note, that the value of $\delta\rho_\Phi/\rho_\Phi \sim \delta\Phi/\Phi$ becomes approximately constant (in time) only at the end of inflation, when the field Φ starts to oscillate about the minimum of $V(\Phi)$. The values of $\delta\Phi(k)$, Φ and $H(\Phi)$ at that time differ somewhat from their values at earlier stages of inflation. Therefore, some care should be taken for an accurate determination of the final shape of the spectrum $\delta\rho_\Phi(k)/\rho_\Phi$. Typically the spectrum of adiabatic perturbations grows slowly (logarithmically) at small k[16,24], see the next section.

---

* One should note that the models do exist in which adiabatic perturbations are generated with a spectrum considerably different from the flat one. For example, in realistic models of inflation several different stages of inflation may exist, which are governed by different inflaton fields Φ_i, see *e.g.* ref. 60. In the intervals which may occur between these stages adiabatic perturbations are not generated. This leads to existence of several different minima in the spectrum. Another possibility appears in the Kaluza-Klein realization of the chaotic inflation scenario[63], in which one may obtain adiabatic perturbations which are strongly suppressed in the long-wave limit.

Eq. (3.13) is valid not only for perturbations of the field Φ, but also for perturbations of other scalar fields φ. However, since $\rho_\Phi \simeq \rho_{\text{tot}} \gg \rho_\varphi$ during inflation (by the definition of the inflaton field Φ), perturbations of all fields except Φ initially do not lead to considerable adiabatic perturbations of ρ_{tot}, and, consequently, to metric and temperature perturbations. Such isothermal perturbations may become important at later stages of the evolution of the universe. If, for example, the field φ is some massive noninteracting field with a mass m, then its potential energy density $V(\varphi)$ starts to decrease only when the Hubble parameter $H(t)$ becomes smaller than m. Then the field φ starts its oscillations about the minimum of $V(\varphi)$, and its energy decreases as the energy density of a nonrelativistic gas, $\rho_\varphi \sim a^{-3}$, whereas the energy density of the hot matter produced by the decay of the inflaton field decreases as a^{-4}. Therefore the value of ρ_φ at later stages of the evolution of the universe may become greater than ρ_Φ, and the isothermal perturbations of ρ_Φ may become important or even dominating. Whereas the time dependence of $H\big(\Phi(t)\big)$, the spectrum of $\delta\Phi(k)$ and the spectrum of adiabatic perturbations $\delta\rho_\varphi(k)$ are determined by the properties of $V(\Phi)$ only, the spectrum of perturbations of other fields φ at a given $H(\Phi)$ is determined by $V(\varphi)$. This makes it rather easy to alter the shape of the spectrum of isothermal perturbations by a proper choice of $V(\Phi)$ and $V(\varphi)$. For example, if during some stage of inflation $V\big(\Phi(t)\big)$ changes slowly, then the spectrum $\delta\rho_\varphi/\rho_\varphi \sim \delta\rho/\rho$ is strongly suppressed in the long-wave region if m_φ^2 is not *much* smaller than $H^2\big(\Phi(t)\big)$, see Eq. (3.13). We will return to this question later, and now we will consider some simple realizations of the mechanism of isothermal density generation discussed above.

4. Generation of Isothermal Perturbations in a Simple Model of Two Noninteracting Scalar Fields

Let us consider a simple model of two noninteracting scalar fields, Φ and φ, with the Lagrangian

$$\mathcal{L} = \frac{1}{2}\left(\partial_\mu \Phi\right)^2 + \frac{1}{2}\left(\partial_\mu \varphi\right)^2 - \frac{1}{4}\lambda_\Phi \Phi^4 - \frac{1}{4}\lambda_\varphi \varphi^4 \tag{4.1}$$

and with $\lambda_\Phi \ll \lambda_\varphi$ [47,48]. Since the only constraint on the initial values of the constant part of the fields Φ and φ is[14,64]

$$V(\Phi) + V(\varphi) = \frac{1}{4}\lambda_\Phi \Phi^4 + \frac{1}{4}\lambda_\varphi \varphi^4 \lesssim M_P^4 , \tag{4.2}$$

one may expect that initially the field Φ was of the order $M_P \lambda^{-1/4}$, and was much larger than the field φ. Moreover, the curvature of the potential $V(\varphi)$ is much greater than the curvature of the potential $V(\Phi)$. Therefore one may expect that the field φ decreases to $\varphi \leq M_P$ much earlier than the field Φ, and the last stages of inflation are driven by the inflaton field Φ:

$$H^2(\Phi) \approx \frac{2\pi\lambda_\Phi \Phi^4}{3M_P^2} , \tag{4.3}$$

where

$$\Phi(t) = \Phi_0 \exp\left(-\sqrt{\frac{\lambda_\Phi}{6\pi}}\, M_P t\right) . \tag{4.4}$$

In this case from the equation for the field φ

$$\ddot{\varphi} + 3H(\Phi)\dot{\varphi} = -\lambda_\varphi \varphi^3 \tag{4.5}$$

it follows that during the last stages of inflation

$$\varphi(t) = \sqrt{\frac{\lambda_\Phi}{\lambda_\varphi}}\, \Phi(t) \tag{4.6}$$

and

$$m^2(\varphi) = m^2(\Phi) = \frac{3M_P\sqrt{3\lambda_\Phi}}{\sqrt{2\pi}}\, H(\Phi)\ , \tag{4.7}$$

where

$$m^2(\varphi) = \frac{d^2V(\varphi)}{d\varphi^2} = 3\lambda_\varphi\varphi^2\ ,\ m^2(\Phi) = \frac{d^2V(\Phi)}{d\Phi^2} = 3\lambda_\Phi\Phi^2\ . \tag{4.8}$$

Thus, at the end of inflation $\Phi(t) \sim M_P$,$\varphi(t) \sim \sqrt{\lambda_\Phi/\lambda_\varphi}M_P$. The amplitudes of perturbations of the fields φ and Φ at that time are equal to each other:

$$\delta\varphi(t) = \delta\Phi(t) \sim \frac{H(t)}{2\pi} \sim \frac{\sqrt{\lambda_\Phi}}{\sqrt{6\pi}M_P}\Phi^2 \sim \sqrt{\lambda_\Phi}M_P\ . \tag{4.9}$$

Perturbations of the field Φ give rise to adiabatic (inflaton) perturbations

$$\frac{\delta\rho_\Phi}{\rho_\Phi} \sim \frac{\delta\Phi}{\Phi} \sim \sqrt{\lambda_\Phi}\ , \tag{4.10}$$

which is a well-known result[16–24]. Perturbations of the field φ give rise to isothermal (isoinflaton) perturbations

$$\frac{\delta\rho_\varphi}{\rho_\varphi} \sim \frac{\delta\varphi}{\varphi} \sim \sqrt{\lambda_\varphi}\ . \tag{4.11}$$

During the first stages of the evolution of the universe after inflation these perturbations are irrelevant, since $\rho_{\rm tot} \simeq \rho_\Phi \gg \rho_\varphi$, and

$$\frac{\delta\rho_\varphi}{\rho_{\rm tot}} \sim \sqrt{\lambda_\Phi}\frac{\sqrt{\lambda_\Phi}}{\lambda_\varphi} \ll \sqrt{\lambda_\Phi}\ . \tag{4.12}$$

Thus, initially the most important density perturbations were the adiabatic ones with an amplitude given by the square root of the *smallest* of the coupling constants λ_φ , λ_Φ (4.10). However, if the field φ weakly interacts with other particles, the main contribution to the total energy density of the universe at later stages of its evolution will be given by the field φ or by its decay products. In this case isothermal perturbations become responsible for the formation of the large-scale structure of the universe, and their magnitude is given by the square root of the *largest* of the coupling constants λ_φ , λ_Φ (4.11). Moreover, since $\lambda_\varphi \gg \lambda_\Phi$, isothermal (isoinflaton) perturbations may play a dominating role even if the energy density of the field φ always remains somewhat smaller than the total energy density $\rho_{\rm tot}$.

Since $m_\varphi^2 = m_\Phi^2$ and $\varphi = \sqrt{\lambda_\Phi/\lambda_\varphi}\,\Phi$ during the last stages of inflation, the spectrum of isothermal perturbations in our model will be similar to that of adiabatic perturbations, *i.e.*, it will be nearly flat. However, this is not a general rule. If, for example, $V(\varphi) = \frac{1}{2}m^2\varphi^2$, and during the last stages of inflation $H(\Phi)$ becomes smaller than m, then the short-wave perturbations $\delta\rho_\varphi(k)$ will be suppressed. On the other hand, if $V(\varphi) = -\frac{1}{2}m^2\varphi^2 + \frac{1}{4}\lambda\varphi^4$ where φ is a complex scalar field, then the Goldstone (massless) mode appears only when the field rolls down to the minimum of $V(\varphi)$. This may occur during the last stages of inflation, and in this case a cut-off in the spectrum $\delta\rho_\varphi(k)$ may appear in the long-wave region. A discussion of other similar possibilities will be contained in the next two sections.

5. Isothermal Perturbations in the Axion Model: An Infrared Cut-off Due to Topological Effects

One of the most popular pretendants to the role of hidden mass of the universe is a classical axion field φ, which is a pseudo-Goldstone boson field with a periodic effective potential of the type

$$V(\varphi) = \alpha\left(1 - \cos\frac{\varphi}{\varphi_0}\right) . \tag{5.1}$$

This means that φ actually is an angle variable. A rather interesting effect which occurs in such theories is a specific damping of the large-scale isothermal perturbations $\delta\rho(k)/\rho$ due to the periodicity of $V(\varphi)$ [47,48,50].

The curvature of $V(\varphi)$ in models of this type typically is extremely small, $m_\varphi^2 = d^2V/d\varphi^2 \sim \alpha/\varphi_0^2 \ll H^2$ during inflation. Therefore fluctuations of the field φ in this model are given by

$$\delta\varphi(k) \sim \frac{H}{2\pi} . \tag{5.2}$$

However, Eq. (3.13) for $\delta\rho_\varphi(k)$ is valid for the theory (5.1) only if the field φ and its dispersion σ are much smaller than φ_0. Let us study inhomogeneities of $V(\varphi)$ induced by the inhomogeneities of the field φ (5.2)[50]. For any given scale l we shall divide the field φ into its large-scale component, which on a scale $l = k^{-1}$ behaves as a constant classical field $\varphi(k)$ and into its short-wave part $\delta\varphi$, corresponding to momenta $k \geq l^{-1}$.

$$\varphi = \varphi(k) + \delta\varphi . \tag{5.3}$$

The value of the "classical" field $\varphi(k)$ is space-dependent (it changes on a distance $l \gg k^{-1}$) and its typical value is given by

$$\tilde{\varphi}(k) = \sqrt{\varphi_c^2 + \sigma_k^2}\,, \tag{5.4}$$

(see eq. (3.10)), where $-\pi\varphi_0 \leq \varphi_c \leq \pi\varphi_0$. For $\sigma_k \gg \varphi_0$ (and this case will be especially important for us) one can write, according to (3.8), (3.11)

$$\tilde{\varphi}^2(k) \sim \sigma_k^2 = \frac{H^2}{4\pi^2} \ln(l_{\max} k)\ . \tag{5.5}$$

The short-wave fluctuations $\delta\varphi$ are distributed in accordance with Gauss' law with dispersion $\tilde{\sigma}_k^2 = \sigma^2 - \sigma_k^2$,

$$\tilde{\sigma}_k^2 = \frac{H^2}{4\pi^2} \ln \frac{H}{k}\ . \tag{5.6}$$

To find the density perturbations produced by fluctuations of the field φ one should consider $V(\varphi)$. Its distribution, the average values $\langle \delta V(k) \rangle$, $\langle \delta V^2(k) \rangle$ etc. on a scale $k \sim l^{-1}$ can be calculated by standard methods. The corresponding perturbations of the axion field energy density are given by

$$\int_k^\infty \delta\rho_\varphi^2(k) \mathrm{d} \ln k = \left\langle\, (\delta V(k))^2 \,\right\rangle - \langle \delta V(k) \rangle^2\ . \tag{5.7}$$

Tedious but straightforward calculations give the following result[50]

$$\frac{\delta\rho_\varphi(k)}{\rho_{\mathrm{tot}}} \approx \frac{\rho_\varphi}{\rho_{\mathrm{tot}}} \sqrt{\beta} \left(\sqrt{2}\cos\theta_k\right) \left(\frac{k}{H}\right)^\beta\,, \tag{5.8}$$

where $\theta_k = \varphi(k)/\varphi_0$, $\beta = \frac{1}{2}(H/2\pi\varphi_0)^2$, $k = l^{-1}$. The two last factors in Eq. (5.8) are rather unusual. The factor $\sqrt{2}\cos\theta_k$ is responsible for the correlation between the values of $\rho_\varphi(k)$ in different domains of the universe. For example, if near some point $\underline{x}$ the total amplitude of all perturbations $\delta\varphi(k)$ with the wavelengths greater than l is given by $2\pi n\varphi_0$, then the density perturbations with this wavelength will be minimal *near this point*, since $\delta V(\varphi)$ near $\varphi = 2\pi n\varphi_0$ is proportional not to $\delta\varphi(k)$ but to $(\delta\varphi(k))^2$. The same effect occurs in the domain near this point in which $\varphi(l)$ (which is a function of $\underline{x}$) remains smaller than $\frac{1}{2}\pi\varphi_0$. The size of such a domain may be very large, and in such a domain a specific correlation exists between those wavelengths at which $\delta\rho(k)$ is minimal and those at which $\delta\rho(k)$ is maximal.

The last factor in (5.8) is especially important. It describes the power-law damping of the long-wave density perturbations. The reason for this effect is that for large l the total dispersion $\delta_k = \sqrt{\langle(\delta\varphi^2)\rangle_k}$ of fluctuations with $l^{-1} < k < H$ becomes greater than $\pi\varphi_0$. This means that inside the domains of the universe of a size $l_0 \gg H^{-1}\exp(2\pi^4\varphi_0^2/H^2)$ any value of the field φ can appear with approximately equal probability. In this case any long-wave fluctuation $\delta\varphi(k)$ with $k^{-1} \sim l \geq l_0$, which is equivalent to a rotation of the distribution of the field φ by an angle $\delta\varphi(k)$, does not change the average distribution of $V(\varphi)$ inside a domain of a size $l \geq l_0$. This explains why the contribution of long-wave perturbations $\delta\varphi(k)$ to the density perturbations in this model is suppressed[47,48,50].

This effect opens some new possibilities in the axion cosmology. One can obtain $\delta\rho_\varphi/\rho_{\text{tot}} \sim 10^{-5}$ on the galaxy scale even if there were no adiabatic perturbations. For example, let us assume that $\varphi_0 \sim 10^{12}$ GeV. In this case, according to ref. 56, the universe should now be axion-dominated, $\rho_\varphi \sim \rho_{\text{tot}}$. If the Hubble parameter H during the last stages of inflation is $H \sim 10^9$ GeV, then from (5.8) it follows that isothermal perturbations are generated with an almost exactly flat spectrum and with $\delta\rho_\varphi/\rho_{\text{tot}} \sim 10^{-5}$. Another possibility is that $\varphi_0 \ll 10^{12}$ GeV, so that the universe is not axion-dominated.

Even in this case it is possible to obtain $\delta\rho_\varphi/\rho_{\text{tot}} \sim 10^{-5}$ if H is not much smaller than φ_0, see (5.8). However, the most interesting possibility is that $\varphi_0 \sim 10^{12}$ GeV, $H \sim 10^{12} - 10^{13}$ GeV. (Note, that such a value of H is of the same order as the results of estimates performed for some simple models of inflation in Sec. 2, see Eqs. (2.10), (2.17).) In this case one can obtain $\delta\rho_\varphi/\rho_{\text{tot}} \sim \delta\rho_\varphi/\rho_\varphi \sim 10^{-5}$ in the axion-dominated universe on a galaxy scale $l_g \geq 10^{22}$ cm for $\beta = \frac{1}{2}(H/2\pi\varphi_0)^2 \sim 0.2$. According to Eq. (5.8), density perturbations of horizon scale $l \sim R_H \sim 10^{28}$ cm will be suppressed by a factor $0\left((l_g/R_H)^\beta\right) \sim 10^{-1}$. However, the suppression is even stronger if one takes into account the time-dependence of the Hubble parameter $H\,((t))$.

For example, in the $\frac{1}{4}\lambda\Phi^4$ theory the value of $H^2(\Phi)$ for which perturbations on a scale $l \sim R_H$ are formed is approximately $\frac{3}{2}$ of the value of H^2 corresponding to perturbations on the galaxy scale. Thus, one may have an effective value of $\beta \sim 0.2$ on galaxy scales and an effective value of $\beta \sim 0.3$ on the horizon scale. This would lead to a suppression of $\delta\rho_\varphi/\rho_{\text{tot}}$ on a scale $l \sim R_H \sim 10^{28}$ cm by a factor $0(3 \times 10^{-2})$ as compared with $\delta\rho_\varphi/\rho_{\text{tot}}$ on a scale $l \sim 10^{22}$ cm. This suppression may be even stronger in models of chaotic inflation, in which $V(\Phi)$ grows at large Φ more rapidly than Φ^4. Therefore the quadrupole anisotropy of the microwave background radiation $(\Delta T/T)_Q$ in the axion-dominated universe may be very small. On the other hand, the small-angle anisotropy $\Delta T/T$ in this model, which corresponds to small-scale density perturbations $\delta\rho_\varphi/\rho_{\text{tot}}$, may be greater than $(\Delta T/T)_Q$.

Let us consider some specific features of the axion dominated universe with isothermal (isoinflation) perturbations (5.8). The universe becomes axion-dominated at a red-shift $Z < Z_{\text{eq}} \sim 10^5$. Isothermal perturbations (5.8) give rise to adiabatic perturbations which at the beginning of the nonlinear stage of large-scale structure formation are given by*

$$\left(\frac{\delta\rho}{\rho}\right)_{\rm ad} = \frac{1}{30}(k\eta)^2 C_i(k)\left(\frac{\delta\rho_\varphi}{\rho_{\rm tot}}\right)_0 , \tag{5.9}$$

where we have denoted initial perturbations (5.8) by $(\delta\rho_\varphi/\rho_{\rm tot})_0$, $\eta = {\rm d}t/a(t)$ is conformal time, $C_i(k)$ is the transfer function for isothermal perturbations. In accordance with the results of numerical calculations, this function is approximately equal to $\left(1+\sqrt{\frac{1}{180}}k\eta_0\right)^{-2}$, where the moment η_0 corresponds to the horizon size $R_0 \sim 234$ Mpc.

Let us consider first the case $\beta \ll 1$, *i.e.* flat isothermal perturbations (5.8), $(\delta\rho_\varphi/\rho_{\rm tot})_0 = \sqrt{\beta}$. The resulting adiabatic perturbations (5.9) will have a flat spectrum somewhat modified by the transfer function $C_i(k)$. This spectrum does not differ considerably from the spectrum arising from initial flat adiabatic perturbations. This is the main difference between the cosmological models with flat isothermal perturbations generated during inflation[47–50] and the previously considered models based on isothermal perturbations with spectra rapidly decreasing at small k.[2]

Let us return now to the discussion of models with isothermal perturbations in which $\delta\rho_\varphi(k)/\rho_{\rm tot}$ is suppressed at small k. We will consider first a toy model* , in which $\delta\rho_\varphi(k)/\rho_{\rm tot}$

---

* One should distinguish between $\delta\rho(k)$, which is normalized on a logarithmic momentum scale and the Fourier-transform $\delta_k = (1/(2\pi)^{3/2})\int {\rm d}^3r e^{ikr}\delta\rho(\underline{r})/\rho$. Those who study inflation usually consider the function $\delta\rho(k)$, whereas in the works on the large-scale structure of the universe the quantity δ_k is commonly used. These quantities are related as follows: $\delta_k = (1/\sqrt{4\pi}k^{3/2})\delta\rho(k)/\rho$.

* This corresponds to the case when the axions appear after some phase transition during inflation[50].

is flat at $k \geq l_0^{-1}$ and vanishes at $k \leq l_0^{-1}$. The multiple components $(\Delta T/T)_{\rm lm}$ of the anisotropy of the microwave background radiation in this case are given by[50]

$$\left(\frac{\Delta T}{T}\right)_{\rm lm} = \frac{5A\sqrt{I_l(l_0)}}{\sqrt{480\pi^2}} , \tag{5.10}$$

where

$$I_l(l_0) = \int_y^{\infty} \frac{{\rm d}k}{k^2} J_{l+1/2}^2(k) , \tag{5.11}$$

$y = 2l_H/l_0$, $l_H \simeq 12000$ Mpc $\sim 10^{28}$ cm (for the present value of $H = 50$ km/Mpc $\cdot$ s). The amplitude of density perturbations A should be approximately 6×10^{-4}, independent of the value of the cut-off $l_0 \geq 200$ Mpc.[50] The coefficient 5, which is absent for initially adiabatic perturbations, when only the Sachs-Wolfe effect works, appears for isothermal perturbation due to the Silk effect** . For a flat spectrum ($l_0 \to \infty$) the quadrupole anisotropy is given by $(\Delta T/T)_Q \simeq 3.5 \times 10^{-5}$.

The integral $I_l(l_0)$ for $l_0 \ll l_H$ is given by $\left(1/(2\pi)^3\right) (l_0/l_H)^2$. Therefore the existence of a cut-off at $l_0 \ll l_H$ leads to a strong suppression of the large-scale perturbations of $\Delta T/T$, and the higher multipoles may be larger than the lower ones.

The presence of the cut-off leads also to a decrease of the galaxy velocity dispersion σ_v by a factor of*** $(1-\sqrt{\frac{1}{20}x})(1+\sqrt{\frac{1}{180}x})^{-3}$, which is equal to 0.6 for $l_0 \sim 0.1 l_H$. This removes one of the main difficulties of cold dark matter models.

** This was pointed out by Starobinsky and Chibisov[58].

*** $x = 2\pi R_0/l_0$.

For the spectrum (5.8) with $\delta\rho_\varphi/\rho \sim \sqrt{\beta}(k/H)^{-\beta}$ instead of the integral $I_l^{1/2}(l_0)$ in (5.10) one should write

$$(5.12) \qquad (Hl_H)^{-\beta}\sqrt{\frac{2^{\beta-2}\Gamma(2-\beta)\Gamma(l+\frac{1}{2}\beta)}{\Gamma^2(\frac{3}{2}-\frac{1}{2}\beta)\Gamma(l+2\frac{1}{2}\beta)}}\,.$$

This quantity should be compared with its value in the case $\beta \ll 1$, where it reduces to $(\pi l(l+1))^{-1/2}$. Taking into account the effective time-dependence of β discussed above, this may lead to a considerable suppression of $(\Delta T/T)_Q$.

In our discussion we considered the axion models with $\varphi_0 \leq 10^{12}$ GeV[56]. However, as has been noted in refs. 72,73, this constraint is not unavoidable in inflationary universe models. Recently this question has become rather important, since a typical value of φ_0 in the superstring theories is $\varphi_0 \sim 10^{16} - 10^{17}$ GeV[74]. Therefore we would like to give a more detailed discussion of this problem here.

The main line of reasoning which leads to such constraints is very simple and apparently irrefutable. At sufficiently small φ (at $\varphi \lesssim \varphi_0$) the effective potential of a weakly interacting scalar field φ usually can be written as $V(\varphi) = \frac{m^2}{2}\varphi^2$. A typical initial value of the field φ in the early universe is $\varphi = 0(\varphi_0)$. During very early stages of the evolution of the universe, when the Hubble parameter H is much bigger than m, the amplitude of the field φ and its energy remain practically unchanged due to the large "friction term" $3H\dot{\varphi}$ in the equation of motion of the field φ. Later, when the value of $H(t)$ becomes smaller than m, the field φ oscillates about the point $\varphi = 0$. Since this field practically does not interact with all other particles and fields, the amplitude of its oscillations decreases solely because of the expansion of the universe: $V(\varphi) \sim a^{-3}(t)$, where $a(t)$ is the scale factor of the universe[78]. Note, that the energy density of massive nonrelativistic particles decreases in the same way, whereas the energy density of a hot dense gas of ultrarelativistic particles decreases much more rapidly, as a^{-4}. For this reason the energy density of the field φ, being originally rather small, may come to dominate at late stages of the evolution of the universe. In particular, in the axion theories with $\varphi \sim 0(\varphi_0) \sim 10^{12}$ GeV the energy density of the oscillating classical axion field at present should be two orders of magnitude greater than the energy density of protons. Since the energy density of protons now constitutes several percent of the overall energy density of the universe $\rho_0 \sim 10^{-29} \mathrm{g}\cdot\mathrm{cm}^{-3}$, it was argued that in the theories with $\varphi_0 \gtrsim 10^{12}$ GeV the energy density of axions at present would be much greater than ρ_0, which would make our universe closed and would cause its premature collapse.

In the inflationary universe scenario the situation looks considerably different. During inflation long-wave perturbations of the axion field φ are generated. As a result, after inflation the universe becomes filled with the field φ gradually changing from $-\varphi_0$ to φ_0 in its different domains. In domains of the size of the same order as the size of the observable part of the universe, $\ell \sim 10^{28}$ cm, the axion field φ typically changes by no more than $0(10)H$, where H is the Hubble parameter at the last stages of inflation. For $H \ll \varphi_0$ this means that our universe becomes divided into many mini-universes of a size $0(10^{28}$ cm) containing an almost homogeneous field φ taking all values from $-\varphi_0$ to φ_0. In principle, we can safely live in a domain with $\varphi \ll \varphi_0$, so that in the observable part of the universe we will have no problems with axions mentioned above[72,73], and no constraints of the type $\varphi_0 \lesssim 10^{12}$ GeV will appear. However, at first glance such a possibility seems very unnatural. Indeed, it seems improbable that we were occasionally born in a non-typical domain of the universe with $\varphi \ll \varphi_0$, and just due to this happy occasion we see no contradiction between the theoretical predictions and the observational data.

The intuitive argument can be made slightly more precise. What we are speaking about is the *conditional probability* for an observer of our type to see the part of the universe with the given properties under an obvious and apparently trivial condition that the observer is a person who is alive during the process of making observations. Now, since all people are made out of baryons (and leptons), let us estimate the fraction of all baryons which exist in those parts of the universe in which the initial value of the field φ happened to be much smaller than φ_0. The process of baryosynthesis in the inflationary universe occurs typically at $T \gtrsim 10^2$ GeV [75–77]. At that time axions still have no effect on the evolution of the universe and the fraction of the volume of the universe in a state with the field in the interval $\varphi_i \lesssim \varphi \lesssim \varphi_i + \Delta\varphi$ was of the order of $\frac{\Delta\varphi}{\varphi_0}$ independent of the value of φ_i. In other words, the probability to find a given value of φ in some part of the universe at a given time was practically independent of φ for $\varphi \lesssim \varphi_0$. Therefore the total number of baryons in domains with $\varphi_i \lesssim \varphi \lesssim \varphi_i + \Delta\varphi$ does not depend on φ, $\frac{N_B(\Delta\varphi)}{N_B(total)} \sim \frac{\Delta\varphi}{\varphi_0}$. For this reason one could argue that most observers, which are made out of baryons, should be located in domains with $\varphi \sim \varphi_0$ rather than with φ close to 0. Thus, a typical observer would see himself not in a safe region with $\varphi \ll \varphi_0$, but in a typical domain of the universe with $\varphi \sim \varphi_0$, and this would contradict the observational data for $\varphi_0 \gg 10^{12}$ GeV.

We see that in the inflationary universe scenario there is *no rigorous proof* of the existence of the constraint $\varphi_0 \lesssim 10^{12}$ GeV, and the derivation of this constraint actually is based on some intuitive anthropic arguments about the plausibility of implausibility of being born in a part of the universe with $\varphi \ll \varphi_0$. Now we are going to show that the arguments suggested above are insufficient to justify the validity of the constraint $\varphi_0 \lesssim 10^{12}$ GeV in the inflationary cosmology.

Indeed, let us consider an "improbable" domain with such

a value of the field $\varphi = \varphi_i \ll \varphi_0$ chosen such that the present energy density ρ_φ of the field φ at present is one or two orders of magnitude greater than the energy density of protons ρ_φ, in agreement with observational data (we will call it "our" domain), and let us consider also another, ten times "more probable" domain, with $\varphi \sim 10\varphi_i \ll \varphi_0$. The total energy density of matter inside this domain at the moment $t = 10^{10}$ years is $\rho_i \sim \rho_\varphi \sim 10^{-29} g \cdot \text{cm}^{-3}$, just as in our domain ($\rho_i = \frac{M_p^2}{6\pi t^2}$ for a flat universe dominated by cold matter or axion fields). However, the *relative* energy density of axions as compared with the density of baryons ρ_φ/ρ_B increases by a factor of $\left(\frac{\varphi}{\varphi_0}\right)^2 \sim 10^2$, which means that the density of protons in this domain is 10^2 times smaller than the density of protons in the first domain. Since the ratio $\frac{n_\gamma}{n_B} = 10^{-9}$ usually does not depend on the value of the axion field, the density of photons in the second domain $n_\gamma \sim T^3$ will also be 10^2 times smaller than in the first one. During the early stages of the evolution of the universe the dominant contribution to the energy density of matter was given by hot relativistic matter. The growth of initial density perturbations $\frac{\delta\rho}{\rho} \sim 10^{-4}$ starts at the moment when the cold matter becomes dominant. As mentioned already this occurs due to the more rapid decrease of $\rho_\gamma \sim T^4 \sim a^{-4}$ as compared with the axion (or proton) density $\rho_\varphi \sim a^{-3}$. In the hot universe the scale factor $a(t)$ is proportional to $\sqrt{t}$. Thus, the second domain will become dominated by cold matter 10^4 times earlier than the first one. The initial density perturbations in both domains will be the same, $\delta\rho/\rho \sim 10^{-5}$, at least if $\frac{H}{\varphi} \lesssim 10^{-4}$ during the last stages of inflation. The rate of growth of perturbations $\delta\rho/\rho$ is the same in both domains. Therefore the time t^*, at which these perturbations grow to $\delta\rho/\rho \sim 0(1)$, in the second domain is 10^4 times smaller than in the first one. The density of galaxies (or galaxy-like objects) at present typically remains the same as at the moment when

$\delta\rho/\rho$ becomes 0(1) and these objects become separated from the general process of the expansion of the universe. Consequently, the average density of matter inside galaxies in the second domain, which is proportional to $\frac{M_p^2}{6\pi(t^*)^2}$, is 10^8 times greater than in our domain.

Thus, we obtained a rather unexpected result. In the second domain, in which the value of the field φ is just one order of magnitude bigger than in our domain, the energy density inside galaxies would be 8 orders of magnitude greater! But it is well known, that the appearance of life of our type requires very special conditions, which in all likelihood are not satisfied in galaxies so radically different from ours. Therefore, even though ten times more baryons do exist in domains with the field φ which is ten times bigger than the field φ_0 in our domain, it seems very plausible that from all these baryons it would be impossible to make one real man (*i.e.* an observer of our type). That is why we see ourselves in a domain filled with a sufficiently small field φ: We just cannot *see* ourselves in domains with large φ.

The argument we have used is based on the so-called anthropic principle, which, roughly speaking, says that a fish sees itself in the water not because the entire universe is filled with water, but because it can only live in water and there is enough water on the earth. Before the invention of the inflationary universe scenario the anthropic principle was considered as being rather metaphysical, since it implied that our universe was created many times before the final success. It was also not clear why the conditions in the observable part of the universe are approximately the same everywhere, whereas it would be easier to make good conditions for life in a much smaller volume. In the context of the inflationary universe scenario this principle, being handled with care, becomes a very useful tool[97]. The inflationary universe itself produces domains of the universe filled with all possible values of the field φ, and the size of each domain filled with an almost homogeneous field φ after inflation becomes much greater than the size of the observable part of the universe. This opens new possibilities in understanding various properties of the part of the universe in which we live now. Simultaneously, this makes it possible to relax or remove some constraints on the elementary particle theories which were obtained in the context of the previous cosmological scheme. In this way we may also obtain an explanation of the small ratio $\frac{\rho_B}{\rho_0} \approx \frac{\rho_B}{\rho_\varphi} \sim 10^{-2}$. Indeed, in our universe there exist domains in which the field φ initially was extremely small, so that at present $\frac{\rho_B}{\rho_\varphi}$ is relatively large, say $\frac{\rho_B}{\rho_\varphi} \sim 0(1)$. However, the number of baryons in such domains is suppressed by a small factor $\left(\frac{\varphi}{f}\right)^2$. On the other hand, there are many more domains with $\varphi \sim \varphi_0$ and $\frac{\rho_B}{\rho_\varphi} \ll 1$. However, in such domains the existence of life of our type is improbable. This gives us the maximal probability to live in a domain in which $\frac{\rho_B}{\rho_0}$ is small, but is not too small. For a certain choice of $\varphi_0 \gg 10^{12}$ GeV it may prove most probable to live in a domain with $\frac{\rho_B}{\rho_0} \sim 10^{-1} - 10^{-2}$. This may make it possible to determine the value of φ_0 for which the observed ratio $\frac{\rho_B}{\rho_0}$ would be the most probable one. However, this would require a much more detailed investigation of galaxy formation, the origin of stars etc., as a function of the value of the field φ.

6. Effects Connected with Phase Transitions During Inflation

In the previous sections we considered theories in which the field φ does not interact with the inflaton field Φ. Meanwhile, such an interaction may lead to some very interesting consequences. As an example, we will consider here a model, which

is a simplest generalization of the model discussed in Sec.4[79]:
(6.1)

$$V(\varphi, \Phi) = \frac{1}{2}m_\Phi^2\Phi^2 + \frac{1}{4}\lambda_\Phi\Phi^4 - \frac{1}{2}m_\varphi^2\varphi^2 + \frac{1}{4}\lambda_\varphi\varphi^4 + \frac{1}{2}\nu\varphi^2\Phi^2 + V_0 .$$

We will assume that $0 < \lambda_\varphi < \nu < \lambda_\varphi$ and that $\lambda_\Phi\lambda_\varphi > \nu^2$ (which is necessary for the global stability of the theory). We will assume also that $m_\Phi^2 < \lambda_\Phi M_P^2$ and that $m_\varphi^2 = c\nu M_P^2$, where $c = 0(1)$. As it was discussed in refs. 14,64, the main constraint on the initial amplitudes of the fields Φ and φ is the inequality $V(\varphi, \Phi) < M_P^4$. Most of the values of Φ and φ satisfying this condition also satisfy the inequality $\Phi > M_P, \Phi > \varphi$. The minimum of $V(\varphi, \Phi)$ for $\Phi > M_P$ is at $\varphi = 0$, and the mass of the field φ at this point is given by

$$m_\varphi^2(\varphi, \Phi) = \nu\Phi^2 - M_\varphi^2 . \tag{6.2}$$

This mass initially is much bigger than the mass of the field Φ and the Hubble constant $H(\varphi, \Phi)$. Therefore the field φ rapidly rolls down to $\varphi = 0$, and inflation is driven by the field Φ. However, when the amplitude of the field Φ decreases to $\Phi < \Phi_c = \frac{M_\varphi}{\sqrt{\nu}}$, the effective mass squared of the field φ becomes negative, and φ starts rolling down to the minimum of its effective potential at

$$\varphi^2 = \frac{M_\varphi^2 - \nu\Phi^2}{\lambda_\varphi} \tag{6.3}$$

Soon thereafter its mass squared again becomes very large and positive. Therefore perturbations in φ are generated only near $\Phi = \Phi_c$. This picks out a particular length for the corresponding density perturbations at the epoch of galaxy formation, $\ell_c \sim \exp\left(\frac{\pi\Phi_c^2}{M_P^2}\right)$ cm. A theory of generation of density perturbations in this model actually is rather complicated, since it involves the simultaneous investigation of perturbations of two scalar fields, $\delta\varphi$ and $\delta\Phi$ together with metric perturbations. This investigation was performed in ref. 79 with the help of computer calculations. The result of this investigation is that the spectrum of adiabatic perturbations in the theory (6.1) acquires a maximum, corresponding to a wavelength slightly different from ℓ_c. Isothermal perturbations in this model also have a maximum, but their amplitude typically is very small.

7. Nonperturbative Effects: Bubbles, Strings, Domain Walls, Monopoles ...

In the previous sections we have discussed small density perturbations generated during inflation. Meanwhile, inflation may lead to a multiple production of exponentially large objects which cannot be completely described by a spectrum of corresponding perturbations. Such objects are bubbles, domain walls, strings, etc., which can be created during inflation. Let us consider several particular examples.

(1) For the last few years there has been great interest in the possibility that the cosmological density perturbations arise due to effects connected with strings[41,42], which are created during high-temperature phase transitions[45] in the early universe (for a review see ref. 43). These strings at the moment of their formation look like Brownian trajectories with a typical persistence length $L \sim T_c^{-1}$, where T_c is the critical temperature of the phase transition. Later the strings tend to straighten out, and their persistence length gradually approaches the size of the horizon $\sim ct$. If the total mass of string per unit length μ is sufficiently large, $\mu \sim (2 \times 10^{16}$ GeV$)^2$, then such strings may give rise to density perturbations $\delta\rho/\rho \sim 10^{-5}$, see refs. 43,46.

However, as we have noted already, thermal production of heavy strings is hardly compatible with inflation. Is there still room for strings with large μ in the inflationary cosmology?

The simplest possibility we see at present is connected with phase transitions which occur during the decrease of the inflaton field Φ in the chaotic inflation scenario. For example, the decrease of the field Φ may cause a phase transition in the model (6.1) if in this model $M_\Phi^2 < 0$, $\nu < 0$. In more complicated models this phase transition, just as the usual high-temperature phase transitions in the expanding and cooling universe, may lead to string formation. The initial persistence length of a typical string would be $L \geq 0(H^{-1})$, and after inflation L becomes exponentially large, $L \sim 0(H^{-1})\exp(4\pi\Phi_c^2/nM_P^2)$, where Φ_c is the critical value of the inflaton field Φ in the theory $V \sim \Phi^n$, at which the phase transition with string formation occurs, see Eqs. (2.9), (2.13). If the phase transition occurs during a later stage of inflation ($\Phi = 0(M_P)$), the density of such strings does not become too small, and they may remain important for cosmology. The distribution of such strings will be considerably different from the distribution of strings produced during the high-temperature phase transition. For example, there will be no strings of a small size. A possible role of such strings in the formation of large-scale structure in the universe requires a separate investigation[80].

The phase transitions caused by the inflaton field Φ can create not only heavy strings, but heavy monopoles as well. The density of these monopoles after inflation is suppressed by the factor $\exp(-12\pi\Phi_c^2/nM_P^2)$, which may be extremely small, so that no primordial monopole problem arises. However, if monopoles are produced near the end of inflation, their density is not totally suppressed, which may encourage those who believe in inflation and still seriously consider the possibility of experimental detection of superheavy monopoles.*

(2) One of the most interesting models in which (light)

---

* The possibility of creation of strings and monopoles during inflation exists also in the new inflationary universe scenario[93,94]; see, however, the comments concerning this scenario in Sec. 2.

strings may appear is the axion theory. Strings are produced in this theory at the moment of the phase transition of the Peccei-Quinn symmetry breaking. In some versions of this theory the phase transition leads also to domain wall formation[81]. This does not contradict cosmology if the Peccei-Quinn phase transition occurs during inflation. There also exist some models in which the domain vacuum structure does not appear at all. However, even in these models at temperatures smaller than $0(10^2)$ MeV there appear domain walls bounded by closed strings[82]. If closed strings are formed due to high-temperature phase transitions, their average size remains small, and they rapidly collapse, radiating their energy away[82]. However, if the Peccei-Quinn symmetry breaking occurs during inflation, the average size of walls bounded by strings becomes exponentially large. Later these "pancakes" either radiate their energy away or form large black holes. A possible role of such objects in the process of formation of large-scale structure in the universe deserves further investigation.*

(3) The first realization of the inflationary universe scenario suggested by Guth[12] was based on the theory of bubble production during first order high-temperature phase transitions[84]. This scenario was not successful because of the production of too large inhomogeneities during expansion and collisions of walls of the bubbles containing a large classical inflaton field Φ [85]. However, if the first-order phase transition occurs not in the inflaton sector, but in the sector of other fields φ, then the inhomogeneities in the total energy density after this phase transition can be reasonably small. If the field φ inside the bubbles is sufficiently homogeneous, the bubbles expand exponentially during inflation. This leads to large-scale inhomogeneities which may form something like a cell structure with maxima of density at the regions where bubble walls collide

---

* For a somewhat different mechanism of formation of exponentially large walls bounded by strings, see ref. 83.

with each other and with minima (voids?) in the central regions of bubbles. Note, that such a picture with exponentially large bubbles appears if the phase transition completes during inflation; otherwise a typical size of a cell would be $0(ct)$, which would be too small.

A typical size of cells formed during inflation at present can be extremely large. It may be that this mechanism is responsible for the appearance of the "wall" of rich clusters $(R \geq 2)$ observed by Kopylov et al[86].

In theories in which first-order phase transitions may proceed simultaneously in many different directions in the isotopic space (which is usually the case in sufficiently complicated theories), the distribution of inhomogeneities may be quite different. For example, during the phase transition in the $SU(5)$ model the $SU(5)$ symmetry in some domains of the universe breaks down to $SU(3) \times SU(2) \times U(1)$, and in some other domains —to $SU(4) \times U(1)$, see *e.g.* ref. 87. If the $SU(5)$ symmetry breaking occurs during inflation, the size of different domains becomes exponentially large. Typically the $SU(4) \times U(1)$ phase is unstable (or metastable), and finally the whole universe may transform to the state $SU(3) \times SU(2) \times U(1)$. However, the energy density (temperature) inside different domains at the same time depends on the *order* of phase transitions inside this domain (directly to $SU(3) \times SU(2) \times U(1)$ or not)** . This leads to the appearance of a domain structure of inhomogeneities in the universe. Typically the $SU(5)$ phase transition proceeds predominantly into one of the phases. In such a case after the final phase

** For example, the difference between the vacuum energy $V(\Phi_1)$ in the $SU(3) \times SU(2) \times U(1)$ phase and the vacuum energy $V(\Phi_2)$ in the $SU(4) \times U(1)$ phase remains constant during the expansion of the universe. Therefore the ratio $\rho_{SU(3)} \cdots / \rho_{SU(4)} \cdots = V(\Phi_1) + C_1T^4 + 0(T^2\Phi^2)/V(\Phi_2) + C_2T^4 + 0(T^2\Phi^2)$ changes in the course of the evolution of the universe.

transition to the $SU(3) \times SU(2) \times U(1)$ phase the universe looks homogeneous everywhere except in some exponentially large islands of a greater (or a smaller) density. The reason is that in the intermediate stage, when some part of space was in a $SU(4) \times U(1)$ phase, the energy density inside each of the phases decreased with a slightly different rate as the universe expanded. The density contrast is model-dependent and can be made as small (or as large) as one wishes. This may lead to the creation of the large-scale structure of the universe without any need for small adiabatic or isothermal density perturbations considered in the previous sections. However, such perturbations may be useful for the creation of substructure inside the islands discussed above, or for gathering these islands into clusters of larger size.

(4) Large domains containing different phases may appear not only due to the first-order phase transitions from one local minimum of $V(\varphi)$ to another deeper minimum. As it follows from the results of Sec. 3, long-wave perturbations of the field φ generated during inflation do not "feel" the curvature of $V(\varphi)$, if $m_\varphi^2 = \mathrm{d}^2V/\mathrm{d}\varphi^2 \ll H^2(\Phi)$. This may lead to phase transitions *from a global minimum of* $V(\varphi)$ *to local minima of* $V(\varphi)$. In particular, this effect may be responsible for the symmetry breaking in SUSY GUTs, in which there typically exist many local minima of $V(\varphi)$ with comparable depth[88]. For a description of such phase transitions see refs. 89,91. As a result, just as in the case considered in the previous subsection, the universe after inflation becomes divided into many exponentially large domains containing matter with different types of symmetry breaking. If finally the phase transition to the most stable phase (to a global minimum of $V(\varphi)$) occurs in all domains (which is necessary to avoid the domain wall problem for domains of a size smaller than the size of the observable part of the universe), then the universe becomes divided into large domains of matter with different energy density inside

each of them.

Thus, there exist many different possibilities to obtain the observed foam-like structure of the universe due to phase transitions during inflation. We wish to note that different mechanisms may lead to a somewhat different structure of the phase(s) inside and outside bubbles. In some versions of our model, baryonic matter is created predominantly in the phase outside bubbles. Density perturbations may exist both in the phase(s) inside and outside bubbles, but only in the phase outside bubbles baryonic matter forms luminous objects. Another possibility is that baryonic matter is present everywhere, but dark matter is present only in the phase outside bubbles and for this reason, density perturbations in the phase inside bubbles are insufficient for galaxy formation. A similar result is true if galaxy formation occurs due to isothermal perturbations of the field ψ, which may be absent inside bubbles. A detailed observational study of the large-scale structure of the universe may show which of these possibilities (if any) is actually realized. This would be important not only for cosmology, but also for the investigation of the phase structure of the underlying theory of elementary particles.

We believe that the nonperturbative effects which occur during inflation, and in particular, the effect of formation of large domains of matter with different energy density inside each of them, can play a very important role in the theory of formation of the large-scale structure of the universe.

8. Conclusions

The development of the inflationary universe scenario[10–14] gives us a possibility of solving simultaneously about ten different problems related to cosmology and elementary particle physics[15]. One of these problems is the problem of the origin of density perturbations necessary for the subsequent formation of the large-scale structure of the universe. As it was shown in refs. 16-24, inflation produces adiabatic density perturbations with an almost flat (scale-free) spectrum. For a long time no other perturbations of a sufficiently large magnitude generated during inflation were known[26]. Therefore the difficulties in constructing the multitude of different objects in the universe from the scale-free adiabatic perturbations became regarded as the difficulties of the inflationary universe scenario. Recently, string models of galaxy formation attracted much attention. For a long time, it remained unnoticed, however, that these models in their original form[41-44], which was based on the theory of high-temperature phase transitions, were hardly compatible with inflation. One could wonder therefore, whether it is possible to suggest any consistent theory of formation of the large-scale structure in the context of the inflationary cosmology.

We do not think at the moment that the diseases of the purely adiabatic theory with flat spectrum are incurable. But in any case it is important to realize that there exist many other previously unexplored possibilities. Some of them are related to isothermal perturbations generated during inflation or to adiabatic perturbations with a non-flat spectrum.

Inflation may lead also to a multiple production of exponentially large strings, walls, bubbles, and other large-scale inhomogeneities of a nonperturbative origin. Investigation of the evolution of such objects may lead to a considerable modification of the usual approach to the problem of formation of the large-scale structure of the universe.

It is also worth noting that a considerable part of the effects discussed in this chapter are directly related to phase transitions which occur during inflation, *i.e.* at an extremely large energy scale. This may give us a new and very interesting possibility for investigating the high-energy behaviour of elementary particle physics and the phase structure of unified theories of elementary particles by a study of the large-scale structure of the universe.

Chapter 2: References

[1] Zeldovich, Ya. B., and Novikov, I. D., *Structure and Evolution of the Universe*, (Nauka, Moscow, 1975) [*Relativistic Astrophysics, Vol. II* (Univ. of Chicago Press, 1983)].

[2] Peebles, P. J. E., *The Large-Scale Structure of the Universe* (Princeton Univ. Press, 1980).

[3] Zeldovich, Ya. B., *Mon. Not. Roy. Astr. Soc.* 160, 1 (1972).

[4] Shandarin, S. F., Doroshkevich, A. G., and Zeldovich, Ya. B., *Sov. Phys. Usp.* 26, 46 (1983).

[5] Grishchuk, L. P., and Zeldovich, Ya. B., *Sov. Phys. A.J.* 22, 125 (1978).

[6] Shandarin, S. F., in *Proc. 117 IAU Symp.* (1985).

[7] Ozernoi, L. M., and Chernin, A. D., *Astron. Zh.* 44, 1131 (1967).

[8] Ostriker, J. P., and Cowie, L. L., *Astrophys. J.* 243, L127 (1980).

[9] Carr, B. J., and Rees, M. J., *Mon. Not. Roy. Astron. Soc.* 206, 801 (1984).

[10] Gliner, E. B., *Sov. Phys. JETP* 22, 378 (1965); *Dokl. Akad. Nauk SSSR* 192, 771 (1970).

[11] Starobinsky, A. A., *Phys. Lett.* 91B, 99 (1980).

[12] Guth, A. H., *Phys. Rev.* D23, 347 (1981).

[13] Linde, A. D., *Phys. Lett.* 108B, 389 (1982); 114B, 431 (1982); 116B, 335 (1982); Albrecht, A., and Steinhardt, P. J., *Phys. Rev. Lett.* 48, 1220 (1982).

[14] Linde, A. D., *Phys. Lett.* 129B, 177 (1983).

[15] Linde, A. D. *Rep. Prog. Phys.* 47, 925 (1984); *Comments Astrophys.* 10, 229 (1985).

[16] Mukhanov, V. F., and Chibisov, G. V., *JETP Lett.* 33, 532 (1981); *Sov. Phys. JETP* 56, 258 (1982).

[17] Hawking, S. W., *Phys. Lett.* 115B, 295 (1982).

[18] Starobinsky, A. A., *Phys. Lett.* 117B, 175 (1982).

[19] Guth, A. H., and Pi, S.-Y., *Phys. Rev. Lett.* 49, 1110 (1982).

[20] Bardeen, J., Steinhardt, P. J., and Turner, M. S., *Phys. Rev.* D28, 679 (1983).

[21] Kodama, H., Tokyo Univ. preprint UTAP-9 (1984).

[22] Kahn, R., and Brandenberger, R., *Phys. Lett.* 141B, 317 (1984).

[23] Hawking, S. W., *Phys. Lett.* 150B, 339 (1985).

[24] Mukhanov, V. F., *Pis'ma ZhETF* 41, 402 (1985).

[25] Turner, M. S., Wilczek, F., and Zee, A., *Phys. Lett.* 120B, 127 (1983); Axenides, M., Brandenberger, R., and Turner, M. S., *Phys. Lett.* 128B, 178 (1983); Steinhardt, P. J., and Turner, M. S., *Phys. Lett.* 129B, 51 (1983).

[26] Brandenberger, R., *Rev. Mod. Phys.* 57, 1 (1985).

[27] Fixen, D. J., Cheng, E. S., and Wilkinson, D. T., *Phys. Rev. Lett.* 50 620 (1983); Ceccarelli, C., Melchiorri, F., Pietranera, L., Dall'Oglio, G., and Melchiorri, B., *Astrophys. J. Lett.* 269, L27 (1983); Strukov, I. A., Sagdeev, R. Z., Kardashev, N. S., Skulachev, D., and Eysmont, N., in *Advances in Space Research, Proc. COSPAR* (Pergamon, 1984).

[28] Rubakov, V. A., Sazhin, M. V., and Veryaskin, A. V., *Phys. Lett.* 115B, 189 (1982).

[29] Starobinsky, A. A., *Sov. Astron. Lett.* 9, 135 (1983); *Pis'ma Astron. Zh.* 11, 323 (1985).

[30] Abbott, L. F., and Wise, M. B., *Phys. Lett.* 135B, 279 (1984).

[31] Lukash, V. N., Naselskij, P. D., and Novikov, I. D., in *Proc. Third Seminar on Quantum Gravity* Eds., Markov, M. A., Berezin, V. A., and Frolov, V. P., (World Scientific, Singapore, 1984).

[32] Klypin, A. A., and Shandarin, S. F., *Mon. Not. Roy. Astr. Soc.* 204, 891 (1983).

[33] Frenk, C. S., White, S. D. M., and Davis, M., *Astrophys. J.* 271, 417 (1983).

[34] White, S. D. M., Frenk, C. S., and Davis, M., *Astrophys. J. Lett.* 274, L1 (1983).

[35] Doroshkevich, A. G., and Khlopov., M. Yu., *Mon. Not. Roy. Astr. Soc.* 211, 277 (1984).

[36] Turner, M. S., Steigman, G., and Krauss, L. M., *Phys. Rev. Lett.* 52, 2090 (1984).

[37] Kofman, L. A., Pogosjan, D. Yu., and Starobinsky, A. A., *Pis'ma Astron. Zh.* 12, 419 (1986).

[38] Davis, M., Efstathiou, G., Frenk, C., and White, S., *Astrophys. J.* 292, 371 (1985); Sahni, V., and Starobinsky, A. A., submitted to Zh. Exp. Teor. Fiz.

[39] Kofman, L. A., and Starobinsky, A. A., *Pis'ma Astron. Zh.* 11, 647 (1985).

[40] Primack, J. R., SLAC preprint PUB-3387 (1984).

[41] Zeldovich, Ya. B., *Mon. Not. Roy. Astro. Soc.* 192, 663 (1980).

[42] Vilenkin, A., *Phys. Rev. Lett.* 46, 1169 (1981).

[43] Vilenkin, A., *Phys. Reports* 121, 263 (1985).

[44] Kibble, T. W. B., *J. Phys.* A9, 1387 (1976).

[45] Kirzhnits, D. A., *JETP Lett.* 15, 529 (1972); Linde, A. D., *Rep. Prog. Phys.* 42, 389 (1979).

[46] Turok, N., and Brandenberger, R., *Phys. Rev.* D33, 2175 (1986).

[47] Linde, A. D., *JETP Lett.* 40, 1333 (1984).

[48] Linde, A. D., *Phys. Lett.* 158B, 375 (1985).

[49] Seckel, D., and Turner, M. S., *Phys. Rev.* 32D, 3178 (1985).

[50] Kofman, L. A., *Phys. Lett.* 173B, 400 (1986).

[51] Peccei, R. D., and Quinn, H., *Phys. Rev. Lett.* 38, 1440 (1977); Weinberg, S., *Phys. Rev. Lett.* 40, 223 (1978); Wilczek, F., *Phys. Rev. Lett.* 40, 279 (1978).

[52] Nilles, H. P., *Phys. Reports* 110, 2 (1984).

[53] Witten, E., *Phys. Lett.* 149B, 351 (1984); 155B, 151 (1985).

[54] Schwarz, J. H., *Phys. Reports* 89, 223 (1982); Green, M., *Surveys in High Energy Phys.* 3, 127 (1983).

[55] Gross., D. J., Harvey, J. A., Martinec, E., and Rohm, R., *Phys. Rev. Lett.* 54, 502 (1985).

[56] Preskill, J., Wise, M. B., and Wilczek, F., *Phys. Lett.* 120B, 127 (1983); Abbott, L. F., and Sikivie, P., *Phys. Lett.* 120B, 133 (1983); Dine, M., and Fischler, W., *Phys. Lett.* 120B, 137 (1983).

[57] Coughlan, G. D., Fischler, W., Kolb, E. W., Raby, S., and Ross, G. G., *Phys. Lett.* 131B, 59 (1983); Goncharov, A. S., Linde, A. D., and Vysotsky, M. I., *Phys. Lett.* 147B, 279 (1984).

[58] Starobinsky, A. A., private communication; Chibisov, G. V., private communication.

[59] Suto, Y., Sato, K., and Sato, H., *Prog. Theor. Phys.* 72, 1137 (1984).

[60] Kofman, L. A., Linde, A. D., and Starobinsky, A. A., *Phys. Lett.* 157B, 361 (1985).

[61] Goncharov, A. S., and Linde, A. D., *Class Quant. Grav.* 1, L75 (1984).

[62] Linde, A. D., *Sov. Phys. JETP* 60, 211 (1984); *Lett. Nuovo Cim.* 39, 401 (1984).

[63] Shafi, Q., and Wetterich, C., *Phys. Lett.* 152B, 51 (1985); Oh, P., *Phys. Lett.* 166B, 292 (1986).

[64] Linde, A. D., *Phys. Lett.* 162B, 281 (1985); *Progr. Theor. Phys. Suppl.* 85, 279 (1985).

[65] Albrecht, A., in *Fifth Workshop on Grand Unification*, eds. Kang, K., Fried, H., and Frampton, P., (World Scientific, Singapore, 1984); Turner, M. S., in *Proc. Cargese School on Fundamental Physics and Cosmology*, eds., Audouze, J., and Tran Thanh Van, J., (Editions Frontieres, Gif-Sur-Yvette, 1985).

[66] Belinsky, V. A., Grishchuk, L. P., Khalatnikov, I. M., and Zeldovich, Ya. B., *ZhETF* 89, 346 (1985).

[67] Piran, T., and Williams, R. M., *Phys. Lett.* 163B, 331 (1985).

[68] Linde, A. D., *Phys. Lett.* 132B, 317 (1983).

[69] Vilenkin, A., and Ford, L. H., *Phys. Rev.* D26, 1231 (1982).

[70] Linde, A. D., *Phys. Lett.* 116B, 335 (1982).

[71] Bunch, T. S., and Davis, P. C. W., *Proc. R. Soc.* A360, 117 (1978).

[72] Pi, S-Y., *Phys. Rev. Lett.* 52, 1725 (1984).

[73] Linde, A. D., Lebedev Physical Institute preprint No. 65 (1985).

[74] Choi, K., and Kim, J. E., *Phys. Lett.* 154B, 393 (1985).

[75] Affleck, I., and Dine, M., *Nucl. Phys.* B249, 361 (1985).

[76] Linde, A. D., *Phys. Lett.* 160B, 243 (1985).

[77] Witten, E., *Nucl. Phys.* B177, 477 (1981).

[78] Turner, M. S., *Phys. Rev.* D28, 1243 (1983).

[79] Kofman, L. A., and Pogosyan, D. Yu., *Phys. Lett.* B214, 508 (1988).

[80] Kofman, L. A., in preparation.

[81] Sikivie, P., *Phys. Rev. Lett.* 48, 1156 (1982).

[82] Kibble, T. W. B., Lazarides, G., and Shafi, Q., *Phys. Rev.* D26, 435 (1982); Vilenkin, A., and Everett, A. E., *Phys. Rev. Lett.* 48, 1867 (1982); Everett, A. E., and Vilenkin, A., *Nucl. Phys.* B207, 43 (1982).

[83] Stecker, F., and Shafi, Q., *Phys. Rev. Lett.* 50, 928 (1983); Lazarides, G., and Shafi, Q., *Phys. Lett.* B148, 35 (1984).

[84] Linde, A. D., *Phys. Lett.* 70B, 306 (1977); *Nucl. Phys.* B216, 421 (1983).

[85] Hawking, S. W., Moss, I. G., Stewart, J. M., *Phys. Rev.* D26, 2681 (1982); Guth, A. H., and Weinberg, E., *Nucl. Phys.* B212, 321 (1983).

[86] Kopylov, A. I., Kuznetsov, D. Yu., Fetisova, D. S., and Schvartzman, V. F., *Astron. Cir.* no. 1347, p. 1 (1984).

[87] Kuzmin, V. A., Shaposhnikov, M. E., and Tkachev, I. I., *Z. Phys.* C12, 83 (1982); Parke, S., and Pi, S-Y., *Phys. Lett.* 107B, 54 (1981); Linde, A. D., Lebedev Physical Institute preprint No. 266 (1981).

[88] Linde, A. D., *Phys. Lett.* 131B, 330 (1983).

[89] Hawking, S. W., and Moss, I. G., *Phys. Lett.* 110B, 35 (1982).

[90] Starobinsky, A. A., in *Fundamental Interactions* (Moscow, MGPI Press, 1984), p. 55.

[91] Goncharov, A. S., and Linde, A. D., *Elem. Chast. Atom. Jad.* 17, 837 (1986), [*Sov. J. Part. Nucl.* 17, 369 (1986)].

[92] Einasto, J., Kofman, L., Starobinsky, A., and Shandarin, S., Acad. Sci. ESSR preprint A-3 (1986).

[93] Shafi, Q., and Vilenkin, V., *Phys. Rev.* D29, 1870 (1984).

[94] Enqvist, K., and Nanopoulos, D. V., *Phys. Lett.* 142B, 349 (1984); *Nucl. Phys.* B252, 357 (1985); Enqvist, K., Kounnas, C., Nanopoulos, D. V., and Quiros, M., *Nucl. Phys.* B262, 538 (1985).

[95] Linde, A. D., *Mod. Phys. Lett.* 1A, 81 (1986); 174B, 395 (1986); *Phys. Scripta* T15, 169 (1987).

[96] Kofman, L. A., Linde, A. D., and Einasto, J., *Nature* 326, 46 (1987).

[97] Linde, A. D., in *300 Years of Gravitation* eds., Hawking. S. W., and Israel, W., (Cambridge University Press, Cambridge, 1987), p. 604.

Chapter 3

QUANTUM COSMOLOGY AND THE STOCHASTIC APPROACH TO INFLATION

1. Inflation and the Wave Function of the Universe

One of the most ambitious approaches to cosmology is based on the investigation of the Wheeler-DeWitt equation for the wave function Ψ of the universe[1,2]. However, this equation has many different solutions, and *a priori* it is not clear which of these solutions describes our universe.

A very interesting idea was suggested by Hartle and Hawking[3]. According to their work, the wave function of the ground state of the universe with a scale factor a filled with a scalar field φ in the semi-classical approximation is given by

$$\Psi_0(a, \varphi) \sim \exp\left(-S_E(a, \varphi)\right) \ . \tag{1.1}$$

Here $S_E(a, \varphi)$ is the Euclidean action corresponding to the

Euclidean solutions of the Lagrange equation for $a(\tau)$ and $\varphi(\tau)$ with the boundary conditions $a(0) = a$, $\varphi(0) = \varphi$. The reason for choosing this particular solution of the Wheeler-DeWitt equation was explained as follows. Let us consider the Green's function of a particle which moves from the point $(0, t')$ to the point $(x, 0)$:

$$
\begin{aligned}
\langle x, 0 | 0, t' \rangle &= \sum_n \Psi_n(x) \Psi_n(0) \exp(i E_n t') \\
&= \int dx(t) \exp\left(iS\left(x(t)\right)\right) ,
\end{aligned}
\tag{1.2}
$$

where Ψ_n is a complete set of energy eigenstates corresponding to energies $E_n \geq 0$. To obtain an expression for the ground-state wave function $\Psi_0(x)$, one should make a rotation $t \to -i\tau$ and take the limit as $\tau' \to -\infty$. In the summation (1.2) only the term $n = 0$ with $E_0 = 0$ survives, and the integral transforms into $\int dx(\tau) \exp\left(-S_E\left(x(\tau)\right)\right)$. Hartle and Hawking have argued that the generalization of this result to the case of interest in the semi-classical approximation would yield eq. (1.1).

The gravitational action corresponding to the Euclidean section S_4 of de Sitter space $\mathrm{d}S_4$ with $a(\tau) = H^{-1}(\varphi) \cos H\tau$ is negative,

$$
\begin{aligned}
S_E(a, \varphi) &= -\frac{1}{2} \int d\eta \left[\left(\frac{\mathrm{d}a}{\mathrm{d}\eta} \right)^2 - a^2 + \frac{\Lambda}{3} a^4 \right] \cdot \\
&\cdot \frac{3\pi M_P^2}{2} = -\frac{3 M_P^4}{16 V(\varphi)} .
\end{aligned}
\tag{1.3}
$$

Here η is the conformal time, $\eta = \int \mathrm{d}t/a(t)$, $\Lambda = 8\pi V/M_P^2$. Therefore, according to (1.1)

$$
\Psi_0(a, \varphi) \sim \exp\left(-S_E(a, \varphi)\right) \sim \exp\left(\frac{3 M_P^4}{16 V(\varphi)} \right) .
\tag{1.4}
$$

This means that the probability P of finding the universe in the state with $\varphi = \text{const}$, $a = H^{-1}(\varphi) = \left(3M_P^2/8\pi V(\varphi)\right)^{1/2}$ is given by

$$P(\varphi) \sim |\Psi_0|^2 \sim \exp\left(\frac{3M_P^4}{8V(\varphi)}\right) . \tag{1.5}$$

This expression has a very sharp maximum as $V(\varphi) \to 0$. Therefore the probability of finding the universe in a state with a large field φ and having a long stage of inflation becomes strongly diminished. One can argue, of course, that at $V(\varphi) \gg M_P^4$ the function $P(\varphi)$ becomes constant, and if one introduces some cut-off at $V(\varphi) \to 0$ and integrates over all values of the field φ from $-\infty$ to $+\infty$, then the probability of finding the universe with large φ (with $V(\varphi) \gg M_P^4$) becomes large[4]. However, this would break the rule according to which one should try not to appeal to processes which occur at densities much greater than the Planck density M_P^4.

There exists an alternate choice of the wave function of the universe. It can be argued that the analogy between the standard theory (1.2) and the gravitational theory (1.3) is incomplete. Indeed, there is an overall minus sign in the expression for $S_E(a, \varphi)$ (1.3), which indicates that the gravitational 'energy' associated with the scale factor a is negative. (This is related to the well-known fact that the total energy of a closed universe is zero, being a sum of the positive energy of matter and the negative energy of the scale factor a). In such a case, to obtain Ψ_0 from (1.2) one should rotate t not to $-i\tau$, but to $+i\tau$, which leads to[5]

$$P \sim |\Psi_0(a, \varphi)|^2 \sim \exp\left(-2|S_E(a, \varphi)|\right) \sim \exp\left(-\frac{3M_P^4}{8V(\varphi)}\right) . \tag{1.6}$$

Actually, this result is valid only if the evolution of the field φ is very slow, so that this field acts only as a cosmological constant $\Lambda(\varphi) = 8\pi V(\varphi)/M_P^2$ in (1.3). Fortunately, this is indeed the case during inflation.

Later the same result was obtained by another method, devised by Zeldovich and Starobinsky[6], Rubakov[7] and Vilenkin[8]. This result can be interpreted as a probability of quantum tunneling of the universe from $a = 0$ (from 'nothing') to $a = H^{-1}(\varphi)$. In complete agreement with the results of the previous section, Eq. (1.6) predicts that a typical initial value of the field φ is given by $V(\varphi) \sim M_P^4$ (if one does not speculate about the possibility that $V(\varphi) \gg M_P^4$), which leads to a very long stage of inflation.

It must be said that there is no *rigorous* derivation of either the Hartle-Hawking wave function, Eq. (1.4), or the tunneling wave function, Eq. (1.6), and the physical meaning of creation of everything from 'nothing' is far from clear. Therefore a deeper understanding of the physical processes in the inflationary universe is necessary in order to investigate the wave function of the universe $\Psi_0(a, \varphi)$ and to suggest a correct interpretation of this wave function. With this purpose we shall try to investigate the global structure of the inflationary universe, and go beyond the minisuperspace approach used in the derivation of Eqs. (1.4) and (1.6).

2. Fluctuating Scalar Field and Eternal Chaotic Inflation

The vacuum structure in the inflationary universe is rather nontrivial[9]. The spectrum of vacuum fluctuations of the scalar field φ with wavelengths $l = k^{-1} \ll H^{-1}$ looks as a usual spectrum of quantum fluctuations in Minkowski space. Meanwhile, the spectrum of fluctuations with wavelengths $l = k^{-1} \gtrsim H^{-1}$ is similar to the spectrum of fluctuations in quantum statistics with anomalously large occupation numbers, $n_k \sim \left(\frac{H}{k}\right)^3 \gg 1$.

In such a case the noncommutativity of creation and annihilation operators becomes unimportant, $[a_k a_k^+] = 1 \ll \langle a_k^+ a_k \rangle = n_k$, and quantum fluctuations with wavelengths $l = k^{-1} \gg H^{-1}$ can be interpreted as inhomogeneities $\delta\varphi(x)$ of the distribution of the *classical* field φ. The theory of this phenomenon is completely analogous to the usual theory of Bose-condensation and of spontaneous symmetry breaking in quantum field theory[10,11].

A detailed investigation of fluctuations of the scalar field φ generated during inflation can be found in refs. 9, 12, 11, and 13. Here we will review some results of this investigation (see also Chapter 2).

After some stages of inflation all original local inhomogeneities of the field $\varphi(x)$ disappear, and the field can be represented as

$$\varphi(x,t) = \varphi_c(t) + \delta\varphi(x,t) \ , \tag{2.1}$$

where $\varphi_c(t)$ is the average classical field satisfying the Klein-Gordon equation in an expanding universe, and $\delta\varphi(x,t)$ are long-wave perturbations of the field φ with the wavelengths $l \gtrsim H^{-1}$. Perturbations $\delta\varphi$ satisfy the following equation of motion

$$\delta\ddot{\varphi} + \frac{3\dot{a}}{a}\delta\dot{\varphi} - \frac{1}{a^2}\Delta(\delta\varphi) = -V''\delta\varphi \ , \tag{2.2}$$

where $a \sim \exp Ht, m^2(\varphi_c) = V'' \equiv \frac{d^2V}{d\varphi^2}$ is the effective mass squared of the field φ. From Eq. (2.2) it follows that the wavelength of any perturbation $\delta\varphi$ grows in time as the scale factor of the universe, $l \sim k^{-1} \sim \exp Ht$. During the stage of inflation $m^2(\varphi) \ll H^2$. For $k \gg H$ the solutions of Eq. (2.2) are the same oscillating functions as in Minkowski space, where $\frac{\dot{a}}{a} = H = 0$. However, when the wavelength of any particular perturbation $\delta\varphi$ becomes greater than H^{-1}, the field $\delta\varphi$ stops its oscillations (due to the large friction term $3H\delta\dot{\varphi}$), and its amplitude becomes almost completely time-independent $\left(\delta\varphi(x,t)\right.$ decreases as $\dot{\varphi}(t)$, *i.e.* very slowly[14]$\left.\right)$. Wavelengths of these frozen perturbations soon become exponentially large, whereas their amplitude remains practically unchanged. This effect alone is responsible for the enchancement of the long-wave vacuum fluctuations with $k \ll H$, which is the reason why the occupation numbers n_k of the vacuum fluctuations with $k \ll H$ are so large and why such fluctuations can be interpreted as long-wave perturbations $\delta\varphi$ of the classical field φ.

The process of generation of the long-wave perturbations $\delta\varphi$ starts at the beginning of inflation. During the quasiexponential expansion of the universe more and more vacuum fluctuations of the scalar field φ are freezing in amplitude when their wavelengths become greater than $0(H^{-1})$. This leads to the permanent process of formation of inhomogeneities $\delta\varphi(x)$ of the classical field φ. The inhomogeneities with wavelengths $l \gtrsim H^{-1}$ formed at some moment t are stretched with the expansion of the universe (their amplitude remaining almost constant), the field distribution becomes (locally) homogeneous again, but in this homogeneous background new inhomogeneities with wavelength $\sim H^{-1}$ are formed, etc. Since the average amplitude and the sign of new perturbations do not depend on the amplitude and the sign of perturbations generated before, the squared dispersion Δ^2 of these perturbations in the exponentially expanding universe grows linearly with time[9],

$$\Delta^2 = \left\langle \delta\varphi^2 \right\rangle = \frac{H^3}{4\pi^2}t \; . \tag{2.3}$$

Roughly speaking, one can say that the field φ at each given point x behaves as a drunken sailor. During each time interval

$\Delta t = H^{-1}$ he makes a step of length

$$|\delta\varphi| = \frac{H}{2\pi} , \tag{2.4}$$

but since the directions of his steps are not correlated, the average distance he manages to travel in time t is given by $\Delta = \sqrt{\langle\delta\varphi^2\rangle} = \frac{H}{2\pi}\sqrt{Ht}$ (see Eq. (2.3)).

Actually, Eq. (2.4) is even more fundamental than Eq. (2.3). Equation (2.4) is valid at each given time interval $\Delta t = H^{-1}$ of inflation, since during this time the value of H remains practically unchanged. However, during the time interval $t \gg H^{-1}$ the value of H may change and the previously generated fluctuations $\delta\varphi$ may also decrease (or increase), which may lead to some corrections to Eq. (2.3)[9]. We will investigate this question in the next section, and now we will consider the evolution of the field φ during several subsequent time intervals $\Delta t = H^{-1}$. According to the Klein-Gordon equation in an expanding universe, in models with $V(\varphi) = \frac{\lambda\varphi^{2n}}{2nM_p^{2n-4}}$ the classical field φ_c during the time $\Delta t = H^{-1}$ decreases by

$$\Delta\varphi = \frac{nM_P^2}{4\pi\varphi} . \tag{2.5}$$

(see Eq. (2.6) of Chapter 2). As follows by comparing Eqs. (2.4) and (2.5), at $\varphi \ll \varphi^*$, where

$$\varphi^* = \lambda^{-1/(2n+2)} M_p , \tag{2.6}$$

the average value of $\delta\varphi(x)$ (2.4) is rather small, $|\delta\varphi(x)| \ll \Delta\varphi(x)$, and therefore fluctuations of the field φ practically do not influence the rolling of the field φ down to the minimum of $V(\varphi)$.

The behavior of the field φ in domains with $\varphi \gg \varphi^*$ is much more complicated and interesting. (Note, that $V(\varphi^*) \sim \lambda^{1/(n+1)} M_P^4 \ll M_P^4$, *i.e.* we are still in the region where the energy density is much smaller than the Planck density.) After the time $\Delta t = H^{-1}$ any domain of initial size $O(H^{-1})$ expands e times, with its volume being increased e^3 times. Therefore this domain can be considered as a combination of $O(e^3)$ domains of a size $\sim H^{-1}$. In almost half of these domains the field φ during this time grows rather than decreases. It is important, that, in accordance with the "no hair" theorem for de Sitter space, the processes inside each domain of a size $l \gtrsim O(H^{-1})$ practically do not depend on the processes in other domains of the universe[15,16]. This means, that after the time $\Delta t = H^{-1}$ any domain of a size $l \gtrsim O\left(H^{-1}(\varphi)\right)$, which contains the field $\varphi \gg \varphi^*$, grows e times and becomes divided into $O(e^3)$ independent domains (mini-universes) of a size $O(H^{-1})$. In almost half of these domains, the field φ increases. As a result, during the time t, in which the value of $H(\varphi)$ changes sufficiently slowly, the overall volume of space filled with a permanently increasing field $\varphi \gg \varphi^*$, grows approximately as $\exp\left((3 - \ln 2)H(\varphi)t\right)$, whereas the volume of the part of the universe in which the field φ does not decrease, grows almost as fast as $\frac{1}{2}\exp\left(3H(\varphi)t\right)$.

This means, first of all, that the evolution of the inflationary universe, in which there was at least one inflationary domain of a size $l \gtrsim H^{-1}(\varphi)$ with $\varphi \gtrsim \varphi^*$, has no end. Such a domain permanently produces new inflationary domains (mini-universes) with $\varphi \gtrsim \varphi^*$, they produce new domains, and this chain reaction proceeds eternally[17–21].

Note that a similar effect may occur in the old and in the new inflationary universe scenario as well. The field φ in the major part of the physical volume may always remain in the unstable state on the top of the effective potential $V(\varphi)$ [22–26]. In the chaotic inflation scenario the result is even more surprising: the field φ in the major part of the physical volume of the universe is permanently climbing the wall towards the most energetically unfavorable values of its potential energy density $V(\varphi)$. This effect leads to especially important modifications of the standard point of view on the global structure of the universe[17–21].

Now let us take into account the growth of the value of $H(\varphi)$ in those domains, in which the field φ *permanently* grows. In such domains the field φ after each time interval $\Delta t = H^{-1}(\varphi)$ grows by $\delta\varphi = \frac{H(\varphi)}{2\pi}$. Therefore this field satisfies the following equation

$$\frac{\partial\varphi}{\partial t} = \frac{H^2(\varphi)}{2\pi} = \frac{4V(\varphi)}{3M_P^2} = \frac{\lambda\varphi^4}{3M_P^2}, \tag{2.7}$$

which yields

$$\varphi_0^{-3} - \varphi^{-3}(t) = \frac{\lambda}{M_P^2} t . \tag{2.8}$$

This means that after the time

$$\tau = \frac{M_P^2}{\lambda\varphi_0^3} \tag{2.9}$$

the field φ becomes infinitely large.

Actually, however, the growth of the field φ may stop at $\varphi \sim \varphi_p \sim \lambda^{-1/2n} M_p$, when $V(\varphi) \sim M_P^4$. The reason is that for $V(\varphi) \sim M_P^4$ the energy density of inhomogeneities $\delta\varphi$, which is proportional to $\partial_\mu(\delta\varphi)\partial^\mu(\delta\varphi) \sim H^4$, becomes greater than $V(\varphi)$. In such domains inflation does not occur. (Note also, that at $V(\varphi) \gg M_P^4$ any classical description of the evolution of the universe becomes impossible). This means that when the field φ approaches $\varphi_p \sim \lambda^{-1/2n} M_p$, the process of self-reproduction of inflationary mini-universes with a growing field φ becomes suppressed. As we have already mentioned, the overall physical volume of all domains with $\varphi \gg \varphi^*$ (2.6), in which the field φ does not decrease, grows almost as fast as $\frac{1}{2}\exp(3H(\varphi)t)$. Since $H(\varphi)$ increases with an increase of φ, the main part of the physical (proper) volume of the universe should be in a state with a maximal possible value of the field φ, *i.e.* with $\varphi \sim \varphi_p \sim \lambda^{-1/2n} M_p$. Note that the results obtained above do not mean that the universe always should remain in a state with $\varphi \sim \varphi_p$ and with the Planck energy density $V(\varphi) \sim M_P^4$. It is necessary to distinguish two different approaches to the problem. In the first case the task is to find what part of initial volume of the universe is in a state with a given field φ. For the solution of this problem it is necessary to find the distribution function $P_c(\varphi, t)$ of the field φ in the comoving coordinate system, or to calculate the magnitude of the average field φ_c and the dispersion Δ of this field in the unit comoving coordinate volume (*i.e.* without taking into account the rate of expansion of the universe which is different for different φ). As will be shown in the next section, the dispersion Δ during inflation often remains smaller than the mean field φ_c. In such a case the mean field φ_c behaves just as a classical field φ. This means that the field φ decreases in the major part of initial (coordinate) volume of the universe in accordance with the results derived by using the classical equations.

In another approach to the problem it is necessary to find what part $P_P(\varphi, t)$ of the physical volume of the universe (*i.e.* the volume which grows with the expansion of the universe) at a given time t is in a state with a value φ. It is this question that was investigated above, and the result is that inflation leads to the permanent growth of volume of those domains of the universe which are filled with a sufficiently large field, $\varphi \gtrsim \varphi^*$, and that the major part of the physical volume of the universe at a given time (in a synchronous coordinate system, see Sec. 4) always remains in a state with $\varphi \sim \varphi_p$,$V(\varphi) \sim M_P^4$. A qualitative discussion of the global structure of the universe in this scenario is given in Chapter 1. A more detailed discussion of this question is contained in subsequent sections.

3. Diffusion of the Scalar Field during Inflation

As we have already written, the magnitude of the field inhomogeneities with wave-lengths $l \gtrsim H^{-1}$ changes during the time interval $\Delta t = H^{-1}$ by $|\delta\varphi(x)| \sim H/2\pi$, whereas the mean field φ_c during the same time decreases by $\Delta\varphi = -\frac{1}{3H}\frac{dV}{d\varphi}\Delta t$. At each point x the evolution of the long-wave component of the field φ looks like a Brownian motion (diffusion) under the action of the external force $-\frac{dV}{d\varphi}$. To describe the Brownian motion of the long-wave classical field φ (*i.e.* of the field with a typical wavelength $l \gtrsim H^{-1}$) one can use the diffusion equation for the probability distribution $P_c(\varphi, t)$. This equation has the following general form[27]

$$\frac{\partial P_c(\varphi,t)}{\partial t} = -\frac{\partial j_c(\varphi,t)}{\partial \varphi} , \tag{3.1}$$

where $j_c(\varphi, t)$ is the probability flux in the (φ, t)-space,

$$j_c(\varphi,t) = -KP_cV' - \frac{\partial}{\partial\varphi}(\mathcal{D}P_c) . \tag{3.2}$$

Here $V' = \frac{dV}{d\varphi}$, $\mathcal{D}$ is the diffusion coefficient and K is the mobility coefficient[27] The values of K and $\mathcal{D}$ in our case can be obtained from the equation for φ_c and from Eq. (2.3)

$$K = \frac{1}{3H} , \tag{3.3}$$

$$\mathcal{D} = \frac{H^3}{8\pi^2} , \tag{3.4}$$

which finally leads to the diffusion equation

$$\frac{\partial P_c}{\partial t} = \frac{\partial}{\partial \varphi}\left(\frac{P_c V'}{3H} + \frac{1}{8\pi^2}\frac{\partial}{\partial \varphi}(H^3 P_c)\right) , \tag{3.5}$$

or, equivalently,

$$\frac{\partial P_c}{\partial t} = \frac{\partial}{\partial \varphi}\left[\frac{1}{3}\left(\frac{3M_P^2}{8\pi V}\right)^{1/2}\left(\frac{8V^2}{3M_P^4}\frac{\partial P_c}{\partial \varphi} + P_c V'\left(1 + \frac{4V}{M_P^4}\right)\right)\right] . \tag{3.6}$$

This equation was first obtained by Starobinsky[12] (see also refs. 13 and 21). (Related methods have been developed also in refs. 26, 28-30.) Before analyzing solutions of this equation, which may help us to look from another point of view at the results obtained in the previous section, let us first make some comments concerning the domain of its validity.

First of all, this equation is valid only during inflation. This means, that in the chaotic inflation scenario this equation is not valid for $\varphi \lesssim M_p$. It is not valid either for $V(\varphi) \gtrsim M_P^4$ (see Sec. 2). Finally, the equation used in the derivation of (3.5), is valid for the mean field φ_c, but in some cases Eq. (2.2) for $\delta\varphi(x)$ should be somewhat improved in order to take into account the modification of the original homogeneous inflationary universe by these fluctuations. This may lead to some modifications of Eq. (3.5). Fortunately, as will be shown in Sec. 4, these modifications do not change most of the results obtained in the present section.

Bearing in mind these constraints on the validity of Eq. (3.6), let us analyze its possible solutions. The first idea which comes into mind is to consider stationary solutions, $\frac{\partial P_c}{\partial t} =$

$0, j_c = j_0$. The most general solution of this type for $V(\varphi) \ll M_P^4$ is given in refs. 12, 18 and 30

$$P_c(\varphi) \sim \text{const}\exp\left(\frac{3M_P^4}{8V}\right) - 2j_0\frac{\sqrt{6\pi V}}{M_p V'}\,. \tag{3.7}$$

In the absence of a stationary probability flux (*i.e.* for $j_0 = 0$), Eq. (3.7) gives the square of the Hartle-Hawking wave function of the universe[3]. If the universe originally was exponentially expanding in some metastable state $\varphi = \varphi_0$ then from (3.7) it follows that the probability of finding a domain of the universe of a size $l \gtrsim H^{-1}$ in a state such that the field φ satisfies $V(\varphi) > V(\varphi_0)$ is proportional to

$$P_c(\varphi) \sim \exp\left(\frac{3M_P^4}{8}\left(\frac{1}{V(\varphi)} - \frac{1}{V(\varphi_0)}\right)\right)\,, \tag{3.7'}$$

where the last term plays the role of the normalization factor in Eq. (3.7)[12]. But this is just the Hawking-Moss equation for the probability of tunneling from φ_0 to φ in the new inflationary universe scenario[16]. Thus, the Hawking-Moss homogeneous tunneling is nothing but local diffusion of the scalar field φ, which looks homogeneous on a scale $l \gtrsim 0(H^{-1})$.

It is important to note, however, that Eq. (3.7) can be valid only during some *intermediate* stage of inflation, when the universe is in a metastable state φ_0 with $V(\varphi_0) \neq 0$. The field φ in the observable part of the universe must eventually roll down to the state with $V(\varphi) = 0$. But in such a case the Hartle-Hawking term in Eq. (3.7) diverges near the minimum of $V(\varphi)$. In other words, this term does not describe a stationary normalizable distribution of probability $P_c(\varphi)$. The physical meaning of this statement can be easily understood if one takes into account that there is *no* diffusion flux which goes upwards from the region with $V(\varphi) = 0$. In particular, there is no diffusion flux from the region $\varphi \lesssim M_P$ to the region $\varphi \gtrsim M_P$ in the chaotic inflation scenario. Therefore, the Hartle-Hawking wave function can be very useful for the description of a metastable vacuum state with $V(\varphi_0) \neq 0$, but at present we do not see how this wave function can be used for a complete description of the inflationary universe.

The situation when including the second term[12] is even more complicated. For $V(\varphi) \ll M_P^4$ the solution is given by

$$P_c(\varphi) = -2j_0 \frac{\sqrt{6\pi V}}{M_P V'} \ . \tag{3.8}$$

Note that in this regime, diffusion can be neglected (the same result can be obtained from (3.5) for $\mathcal{D} = 0$). Thus, this solution describes classical rolling of the field φ down to $\varphi = 0$. The corresponding distribution $P_c(\varphi, t)$ can be stationary only if there exists a constant probability flux j_0 which goes from the region $V(\varphi) \gg M_P^4$, where Eqs. (3.5) and (3.6) are not valid. Formally, this solution does not exist at all, since it changes its sign when one goes from $\varphi > 0$ to $\varphi < 0$, whereas the probability P_c must be positive. One may try to make some sense out of Eq. (3.8) by considering it separately for $\varphi \gtrsim M_P$ and for $\varphi \lesssim -M_P$, and by interpreting the flux j_0 as a result of quantum creation of universes at $V(\varphi) \gtrsim M_P^4$ [12]. However, the validity of such an interpretation is not quite clear to us. Moreover, it is not clear why such a flux should be stationary, since the process as a whole is not stationary. For example, the total coordinate volume, in which the field φ has rolled down to $\varphi = 0$, grows permanently. In any case, as it was pointed out in ref. 12, (where Eq. (3.7) was first obtained), at present we cannot prove the existence of this solution and say anything definite about the value of j_0 if it is nonzero. Therefore, in what follows we will investigate general nonstationary solutions of the diffusion equation (3.6), which have been obtained in refs. 18-21. This will help us look from a more general point of view at various solutions of this equation, and to confirm the results obtained in the previous section. But before solving this equation we will try to obtain a qualitative understanding of the time evolution of the probability distribution $P_c(\varphi, t)$ in the theory $V(\varphi) = \frac{\lambda}{4}\varphi^4$.

Let us remember that the initial inhomogeneities of the classical scalar field $\varphi(x)$ in the inflationary universe effectively disappear after a time Δt of the order H^{-1} (their wavelengths increase $\exp(H\Delta t)$ times), whereas the new inhomogeneities generated by quantum fluctuations have wavelengths $l \gtrsim H^{-1}$. As a result, on a scale $l \lesssim H^{-1}$ the field φ always remains almost homogeneous. To be more precise, in a domain of a size $l \sim H^{-1}$ a typical deviation of the field $\varphi(x)$ from its average value φ_c inside this domain is given by $|\delta\varphi| \sim \frac{H}{2\pi}$ (2.4). In all theories of interest $H \ll \varphi$ during inflation, and therefore one can take the condition $P_c(\varphi, t = 0) = \delta(\varphi - \varphi_0)$ as a fairly general initial condition for the distribution $P_c(\varphi, t)$ in a domain of initial size $0(H^{-1})$.* Since the evolution of $P_c(\varphi, t)$ in different domains of initial size $\Delta l \sim H^{-1}$ proceeds independently (due to the "no-hair" theorem for de Sitter space), an investigation of the behavior of $P_c(\varphi, t)$ in one domain of initial size $0(H^{-1})$ is quite sufficient to obtain a general picture of the evolution of the inflationary universe. Thus, in what follows we will consider a domain of initial size $0(H^{-1})$ in which initially $P_c(\varphi, t = 0) = \delta(\varphi - \varphi_0)$. The subsequent evolution of the distribution $P_c(\varphi, t)$ in the theory with $V(\varphi) = \frac{\lambda}{4}\varphi^4$ can be divided into two basic stages.

During the first stage, which has a duration $\Delta t \sim \left(\sqrt{\lambda}M_P\right)^{-1}$, the classical field φ remains almost unchanged, $\varphi_c \approx \varphi_0$ (see Eq. (2.8)of Chapter 2). Meanwhile, on the background $\varphi \approx \varphi_0$

---

* Note that any dependence of the behavior of $\varphi(t)$ on the initial value of $\dot{\varphi}(t)$ disappears after the time $\Delta t \gtrsim H^{-1}$ (see for example, ref. 31).

the inhomogeneities $\delta\varphi(x)$ are created, which have dispersion $\Delta = \frac{H}{2\pi}\sqrt{Ht}$ (2.3). At the end of the first stage the dispersion grows to

$$\Delta_0 = \frac{AH}{2\pi}\sqrt{\frac{H}{\sqrt{\lambda}M_P}} = A\frac{\sqrt{\lambda}\varphi_0^3}{M_P^2}\,, \qquad A = 0(1)\,. \tag{3.9}$$

From (3.9) it follows that for $V(\varphi_0) = \frac{\lambda\varphi_0^4}{4} \ll M_P^4$ the dispersion of inhomogeneities $\delta\varphi(x)$ formed inside the domain of initial size $0(H^{-1})$ during the first stage is negligibly small as compared to the average value of the field φ inside this domain, *i.e.* $\varphi \approx \varphi_c \approx \varphi_0$.

During the second stage $\left(t > \left(\sqrt{\lambda}M_P\right)^{-1}\right)$ the mean field φ_c decreases exponentially. Fluctuations of the field φ are also generated, but their amplitude, which is proportional to $H(\varphi)$, decreases as $\varphi^2(t)/\varphi_0^2$, and their dispersion (3.9) decreases as $\varphi^3(t)/\varphi_0^3$. Meanwhile, as was shown in ref. 14, the amplitude of previously created inhomogeneities decreases only as $\dot{\varphi} \sim \varphi(t)$. The reason is that $\dot{\varphi}(x,t)$ obeys the same equation (2.2) as the perturbations $\delta\varphi(x,t)$, and asymptotically (after the time $\Delta t \sim 0(H^{-1})$) they approach the same regime, so that $\frac{\Delta(t)}{\dot{\varphi}(t)} \sim \frac{\delta\varphi(t)}{\dot{\varphi}(t)} = \text{const}$. The dispersion $\Delta(t)$ of inhomogeneities generated during the first stage decreases in the same way, *i.e.*

$$\Delta(t) \sim \Delta_0\frac{\varphi(t)}{\varphi_0} \sim \frac{\sqrt{\lambda}\varphi(t)\varphi_0^2}{M_P^2}A\,. \tag{3.10}$$

This means that the form of the distribution $P_c(\varphi,t)$ is determined by the fluctuations produced during the first stage of the process

(3.11)

$$P_c(\varphi,t) \sim \exp\left\{-\frac{(\varphi-\varphi_c(t))^2}{2\Delta^2(t)}\right\} \sim \exp\left\{-\frac{M_P^4(\varphi-\varphi_c(t))^2}{2A^2\lambda\varphi^2(t)\varphi_0^4}\right\},$$

where $\varphi_c(t)$ is the average field. (Note that the dispersion $\Delta(t)$

(3.10) for $V(\varphi) \ll M_P^4$ always is much smaller than $\varphi_c(t)$). This is why one can use the classical equation of motion and its solution to investigate the evolution of the average field $\varphi_c(t)$. For some potentials $V(\varphi)$ the condition $\Delta \ll \varphi_c$ can be violated at late stages of the evolution of the field φ_c. In such a case an investigation of the evolution of the average field φ_c becomes more complicated).

Instead of the distribution $P_c(\varphi, t)$ in the comoving reference frame (3.11), one may consider the distribution $P_p(\varphi, t)$, taking into account the more rapid growth of domains filled with a large field φ. The relation between P_c and P_p in general is rather complicated. However, for our purposes it is sufficient to know the relation between $P_p(\varphi, t)$ and $P_c(\varphi, t)$ which is approximately correct for $\varphi - \varphi_c \gg \Delta(t)$ after the time $\Delta t \sim \left(\sqrt{\lambda} M_P\right)^{-1}$ from the beginning of the diffusion process. This relation is rather simple, since during the time Δt fluctuations $\delta\varphi$ *typically* change the value of φ by no more than $0(\Delta)$. Therefore, if the field φ in some domain increased in such a way that $\varphi - \varphi_c(t) \gg \Delta(t)$, then during the time Δt it remains approximately at the same distance from the field $\varphi_c(t)$. During this time the physical volume of such a domain grows by $\exp(3H(\varphi)\Delta t)$, whereas the average growth of volume of all domains is given by $\exp(3H(\varphi_c)\Delta t)$. Thus, for $\varphi - \varphi_c \gg \Delta(t)$,

$$P_p(\varphi, \Delta t) \sim P_c(\varphi, t) \exp\left(3\Delta t\left(H(\varphi) - H(\varphi_c)\right)\right)$$

$$\sim \exp\left(-\frac{(\varphi - \varphi_c)^2 M_P^4}{2A^2\lambda\varphi_c^2\varphi_0^4} + \frac{3B(\varphi^2 - \varphi_c^2)}{M_P^2}\right), \tag{3.12}$$

where $A, B = 0(1)$. From (3.12) it follows that in the regions with $\varphi_0 \gtrsim \varphi^* = \lambda^{-1/6} M_P$ during the time $\Delta t \sim \left(\sqrt{\lambda} M_P\right)^{-1}$ the delta-functional distribution $P_p(\varphi, 0) = P_c(\varphi, 0) = \delta(\varphi - \varphi_0)$ transforms into the distribution which grows as φ increases. This conclusion is in agreement with the results obtained in the previous section.

After this preliminary investigation, which clarifies some features of the solutions to be obtained, let us solve Eq. (3.6) for the theories with $V(\varphi) = \frac{\lambda \varphi^{2n}}{2n M_P^{2n-4}}$ with the initial condition $P_c(\varphi, 0) = \delta(\varphi - \varphi_0)$. It is convenient to introduce the dimensionless variables

$$\tau = \frac{M_P}{4\pi} t \ , \quad x = \frac{\varphi}{M_P} \ , \quad h = \frac{H}{M_P} \ . \tag{3.13}$$

In these variables, Eq. (4.5) takes the following form

$$\frac{\partial}{\partial \tau} P_c = \frac{\partial}{\partial x} \left(-A(x) P_c\right) + \frac{1}{2} \frac{\partial^2}{\partial x^2} \left(B(x) P_c\right) \ , \tag{3.14}$$

where $A(x) = \frac{\partial h}{\partial x}$, $B(x) = \frac{h^3}{\pi}$. The relative roles of the first ("convection") and the second ("diffusion") terms in (3.14) can be estimated by the ratio

$$|A(x)| \Big/ \frac{\partial}{\partial x} B(x) \sim h^{-2}(x) \ .$$

Thus, the diffusion term is relatively large only for extremely large values of the field φ, such that $H(\varphi) \sim M_P$, $V(\varphi) \sim M_P^4$. We will consider two possibilities:

(i) $V(\varphi) \ll M_P^4$ and

(ii) $V(\varphi) \sim M_P^4$.

In the first case, in accordance with the smallness of the diffusion term in (3.14) one can expect (see also our previous investigation) that the distribution $P_c(\varphi, t)$ has a sharp peak moving towards the minimum of the potential $V(\varphi)$. The width of this peak will depend on the location of the maximum of $P_c(\varphi)$. To investigate the behavior of $P_c(\varphi, t) = P_c(x, t)$, let us introduce the function $x_c(t)$, which satisfies the differential equation $\dot{x}_c = A(x_c)$ with the initial condition $x_c = x_0 = \frac{\varphi_0}{M_P}$. Let us make another change of variables, $(x, \tau) \to (y, s)$:

$$s = \tau \ , \qquad y = x - x_c(\tau) \ .$$

After this transformation, $\frac{\partial}{\partial \tau} = \frac{\partial}{\partial s} - \dot{x}_c \frac{\partial}{\partial y} = \frac{\partial}{\partial s} - A(x_c)\frac{\partial}{\partial y}$, $\frac{\partial}{\partial x} = \frac{\partial}{\partial y}$ and Eq. (3.14) takes the form
(3.15)

$$\frac{\partial}{\partial s} P_c = -\frac{\partial}{\partial y} \left\{ \left[A(x_c + y) - A(x_c) \right] P_c \right\} + \frac{1}{2} \frac{\partial^2}{\partial y^2} \left[B(x_c + y) P_c \right] \ .$$

The function $x_c(t)$ gives the location of the maximum of $P_c(x)$, whereas y is a measure of the deviation from this maximum. The narrowness of the peak in the distribution $P_c(x)$ makes it possible to use, instead of $A(x_c + y)$ and $B(x_c + y)$, several leading terms of their expansion in the small parameter y. After that, Eq. (3.15) takes the following form

$$\frac{\partial}{\partial s} P_c = -A'(x_c) \frac{\partial}{\partial y}(y P_c) + \frac{1}{2} B(x_c) \frac{\partial^2}{\partial y^2} P_c \ . \tag{3.16}$$

The solution of Eq. (3.16) with the initial condition $P(y, 0) = \delta(y)$ is given by the Gaussian distribution with $\langle y(s) \rangle = 0$ and with the time-dependent dispersion $\Delta(s)$. The value of dispersion can be determined without solving Eq. (3.16) explicitly. Indeed, from (3.16) and from the definition of dispersion

$$\Delta^2(s) = \langle y^2(s) \rangle = \int dy \; y^2 P_c(y, s) \ , \tag{3.17}$$

it follows that $\Delta(s)$ obeys the equation

$$\frac{d}{ds} \Delta^2 = 2A'\left(x_c(s)\right) \Delta^2(s) + B\left(x_c(s)\right) \ . \tag{3.18}$$

Since we are actually interested not in the time evolution of Δ, but in the dependence of the dispersion on the position of the maximum $x_c(s)$ of the distribution P_c, it is convenient to re-write Eq. (3.18) in the following form

$$\frac{d}{dx_c}\Delta^2 = \frac{1}{\dot{x}_c}\frac{d}{ds}\Delta^2 = 2\frac{A'(x_c)}{A(x_c)}\Delta^2 + \frac{B(x_c)}{A(x_c)} . \tag{3.19}$$

The solution of Eq. (3.19) with the boundary condition $\Delta\,(x_c(0)) = 0$ is given by

$$\Delta^2(x_c) = A^2(x_c)\int_{x_0}^{x_c}\frac{B(z)}{A^3(z)}dz . \tag{3.20}$$

In theories with the effective potentials $V(\varphi) = \frac{\lambda\varphi^{2n}}{2nM_P^{2n-4}}$ (in which $h(x) = 2\sqrt{\frac{\pi\lambda}{3n}x^n}$)

$$\Delta^2 = \frac{\lambda x_c^{2n-2}}{3n^2}(x_0^4 - x_c^4) = \frac{\lambda\varphi_c^{2n-2}}{3n^2M_P^{2n}}(\varphi_0^4 - \varphi_c^4) . \tag{3.21}$$

From Eq. (3.21) it follows that the dispersion at $x_c = x_0$ is equal to zero, as it should be. As x decreases, the dispersion (3.21) starts to grow in accordance with Eq. (2.3) (the first stage). In theories with $n > 1$ this growth stops when x_c becomes considerably smaller than x_0, and then $\Delta^2(x_c)$ behaves as x_c^{2n-2} (the second stage), *i.e.* it decreases for $n > 1$, increases for $n < 1$ and becomes constant for $n = 1$. The distribution function $P_c(\varphi, t)$ in the vicinity of its maximum is given by (3.22)

$$P_c(\varphi,t) \sim \exp\left(-\frac{(\varphi-\varphi_c)^2}{2\Delta^2}\right) = \exp\left(-\frac{3n^2(\varphi-\varphi_c)^2M_P^{2n}}{2\lambda\varphi_c^{2n-2}(\varphi_0^4-\varphi_c^4)}\right) .$$

After the end of the first stage (*i.e.* after φ_c^4 becomes much smaller than φ_0^4)

$$P_c(\varphi,t) \sim \exp\left(-\frac{3n^2(\varphi-\varphi_c)^2 M_P^{2n}}{2\lambda\varphi_c^{2n-2}\varphi_0^4}\right) , \qquad (3.23)$$

which is in agreement with our previous estimate (3.11) for the theory $V(\varphi) = \frac{\lambda\varphi^4}{4}$. Consequently, one can again obtain Eq. (3.12) for the distribution $P_p(\varphi, \Delta t)$ at the end of the first stage.

These results are valid when the difference between φ and $\varphi_c(t)$ is sufficiently small, so that it is possible to keep only the first terms of the expansion of A and B in the small parameter y. In the theory $\frac{\lambda}{4}\varphi^4$ this condition is satisfied for $y \lesssim \frac{1}{6}x_c$, *i.e.* for $\varphi - \varphi_c \lesssim \frac{1}{6}\varphi_c$. From (3.21) it follows that in this theory for $\varphi_c \sim \varphi_0$

$$\frac{\Delta(\varphi_c)}{\varphi_c} \lesssim \frac{V(\varphi_0)}{M_P^4} . \qquad (3.24)$$

Therefore for $V(\varphi_0) \ll M_P^4$ the peak of the distribution $P_c(\varphi,t)$ is very narrow (for other theories this is also true, at least if φ_c is not much smaller than φ_0 and the distribution near the peak is actually Gaussian, Eq. (3.23)). However, for $\varphi \gg \varphi_0$, for $\varphi \ll \varphi_0$ and if initially $V(\varphi_0) \sim M_P^4$, the corresponding distribution is not Gaussian.

An investigation of diffusion starting with $V(\varphi_0) \sim M_P^4$ may be of some interest since it may be connected with the process of creation of the universe from space-time foam (from "nothing"). Simultaneously, we will obtain a more detailed description of the distribution $P_c(\varphi,t)$ in the first stage of the diffusion process considered above (with $V(\varphi_0) \ll M_P^4$), which will be valid outside the narrow peak near $\varphi_c(t)$. Since this approach will be valid for $\varphi - \varphi_c(t) \gg \Delta$, the distribution function $P_c(\varphi,t)$ can be represented as $\exp(-S(\varphi,t))$ where

$S(\varphi, t) \gg 1$. In such a case the functions $A(x)$ and $B(x)$ in (3.14) change with x much more slowly than $P_c(x)$, the second (diffusion) term in (3.14) plays the leading role, and Eq. (3.14) can be re-written as follows:

$$\frac{\partial}{\partial \tau} P_c = \sqrt{\frac{B(x)}{2}} \frac{\partial}{\partial x} \left[\sqrt{\frac{B(x)}{2}} \frac{\partial P_c}{\partial x} \right] . \tag{3.25}$$

After the change of variables $\xi = \int \frac{dx}{\sqrt{\frac{B(x)}{2}}}$ we obtain the simplest diffusion equation

$$\frac{\partial}{\partial \tau} P_c = \frac{\partial^2}{\partial \xi^2} P_c . \tag{3.26}$$

The solution of this equation, which at $\tau = 0$ differs from zero only at the point $x = x_0$, is proportional to

$$\exp \left\{ - \frac{\left(\xi - \xi(x_0) \right)^2}{4\tau} \right\} . \tag{3.27}$$

In the theory $V(\varphi) = \frac{\lambda \varphi^{2n}}{2n M_P^{2n-4}}$

$$\xi = - \frac{1}{3n-2} \left(\frac{3n\sqrt{n}}{\lambda\sqrt{\lambda\pi}} \right)^{1/2} \frac{1}{x^{(3n/2)-1}}$$

and the distribution $P_c(x)$ is proportional to

$$P_c(x) \sim \exp \left\{ - \frac{3n\sqrt{n}}{4\lambda\sqrt{\lambda\pi}} \cdot \frac{1}{(3n-2)^2 \tau} \left[\frac{1}{x^{(3n/2)-1}} - \frac{1}{x_0^{(3n/2)-1}} \right]^2 \right\} . \tag{3.28}$$

Note, that for $x \ll x_0$ this solution does not depend on x_0

$$P_c(x) \sim \exp \left\{ - \frac{3n\sqrt{n}}{4\lambda\sqrt{\lambda\pi}(3n-2)^2 \tau x^{3n-2}} \right\} . \tag{3.29}$$

This result was obtained after neglecting the convective term in Eq. (3.14) as compared with the diffusion term. Convection (classical motion) can actually be neglected for sufficiently small time intervals, τ, during which the classical change of the initial value of x is much smaller than the change that occurs due to the Brownian motion. Let us estimate the duration of the time interval τ, during which the decrease of the coordinate x (of the field φ) occurs mainly due to diffusion rather than to the classical motion. The speed of the classical motion is determined by the equation

$$\dot{x} = A(x) = -\frac{\partial h}{\partial x} = -2\sqrt{\frac{\pi n \lambda}{3}} x^{n-1} , \tag{3.30}$$

whereas the speed of diffusion near the point x can be estimated by the ratio $\frac{x}{\tau}$. Indeed, according to (3.29), the coordinate $x(t)$ corresponding to $P_c\big(x(\tau)\big) = \text{const}$, moves with speed $\sim -(3n-2)\frac{x}{\tau}$. This means that diffusion dominates for $\tau \lesssim \lambda^{-1/2} x^{2-n}$. At the end of this stage in the theories with $n \leq 2$

$$P_c(x) \sim \exp\left\{-\frac{c}{\lambda x^{2n}}\right\} , \quad c = 0(1) , \tag{3.31}$$

i.e.

$$P_c(\varphi) \sim \exp\left\{-\frac{\bar{c} M_P^4}{V(\varphi)}\right\} , \quad \bar{c} = 0(1) \tag{3.32}$$

(for $\varphi \ll \varphi_0$). The physical meaning of this result is very interesting. The distribution $P(\varphi)$ (3.32) gives the probability of quantum creation of an inflationary domain (mini-universe) of a size $l \gtrsim H^{-1}(\varphi)$, containing the classical field φ. This process occurs due to the Brownian motion of the fluctuating field φ in the interior of some inflationary domain with $\varphi_0 > \varphi$. Note, that $P_c(\varphi)$ does not depend on the value of φ_0 for $\varphi \ll \varphi_0$ (see Eqs. (3.28) and (3.29)). Since the probability of existence of an inflationary domain of size $0(M_P^{-1})$

with $V(\varphi_0) \sim M_P^4$ in the space-time foam is not exponentially suppressed[5–8,32], Eq. (3.32) gives the probability of quantum creation of the inflationary universe from the space-time foam (from "nothing") [18–21]. This result is in agreement with the previous estimates of the probability of quantum creation of the universe, which were based on the improved calculation of the wave function of the universe[5] and on the idea about tunneling "from nothing"[6–8].

Another interesting result which follows from Eq. (3.28) is that the distribution $P_c(\varphi, t)$ for $\varphi \gg \varphi_0$ does not depend on φ

$$P_c(\varphi, t) \sim \exp\left(-\frac{\alpha}{t}\right) , \tag{3.33}$$

where

$$\alpha = \frac{3n\sqrt{n}}{4(3n-2)^2 M_P \sqrt{\lambda\pi}} \left(\frac{M_P}{\varphi_0}\right)^{3n-2} . \tag{3.34}$$

This means that the distribution $P_p(\varphi, t)$ for $\varphi \gg \varphi_0$ grows even more rapidly than for $\Delta \ll \varphi - \varphi_0 \lesssim \frac{\varphi_0}{6}$, Eq. (3.12).

4. Fluctuating Field in a Fluctuating Metric

In our previous investigation we have implicitly used the system of coordinates in which the local rate of the expansion of the universe was determined by the value of the Hubble constant $H(\varphi) \sim \sqrt{V(\varphi)}$ at a given point x

$$ds^2 \approx dt^2 - a^2(x,t)d\underline{x}^2 , \tag{4.1}$$

where

$$a(x,t) \sim \exp\left(\int_0^t H\left(\varphi(x,t)\right) dt\right) . \tag{4.2}$$

However, we have not analyzed to which extent this synchronous coordinate system can be used and how our results look in other coordinate systems. In other words, we would like to have a theory which describes simultaneously fluctuations of the scalar field φ and fluctuations of the metric.

The basic features of this theory were elaborated in 1981-1982 and applied to the Starobinsky model[33] and to the new inflationary universe scenario[14]. In the chaotic inflation scenario the corresponding theory is slightly more complicated due to a more considerable backreaction of metric perturbations on perturbations of the scalar field $\delta\varphi$ (see below). A self-consistent approach to this problem was suggested in ref. 34. In what follows we will use the methods developed in ref. 34 to investigate the global structure of the inflationary universe.

We will use the conformally-Newtonian coordinate system[34,35]

$$(4.3) \qquad ds^2 = (1 + 2\Phi(x,\tau))\, d\tau^2 - (1 - 2\Psi(x,\tau))\, a^2(\tau) d\underline{x}^2 \ .$$

The functions Φ and Ψ are gauge-invariant quantities, coinciding with the functions Φ_H and Φ_A introduced by Bardeen[36]. It is important that these gauge-invariant functions describe metric perturbations in the coordinate system (4.3).

As before, we will consider a domain of the inflationary universe of initial size $l \sim 0(H^{-1}(\varphi))$ which contains a sufficiently homogeneous scalar field φ. All initial inhomogeneities of this field soon disappear due to inflation, and the field φ in this domain can be represented as a sum (2.1)

$$(4.4) \qquad \varphi(x,\tau) = \varphi_c(\tau) + \delta\varphi(x,\tau) \ ,$$

where the inhomogeneities $\delta\varphi(x,\tau)$ appear due to the generation of the long-wave perturbations (2.1) and (2.2). In particular, the initial value of $\delta\varphi$ in a domain of a size $0(H^{-1})$ can be estimated by $|\delta\varphi| \sim \frac{H}{2\pi}$ (2.4). Hence, $\delta\varphi/\varphi \ll 1$.

The leading contribution to $\delta\varphi(x,t)$ after some period of inflation is given by fluctuations with exponentially large wavelengths. The main contribution to the long-wave inhomogeneities of the energy-momentum tensor T^ν_μ is given not by terms proportional to spatial gradients of $\delta\varphi(x,\tau)$, but by terms proportional either to $(\partial_0(\delta\varphi))^2$ $,\partial_0\varphi \cdot \partial_0\left(\delta\varphi(x,t)\right)$, or to $\delta V(\varphi) \sim \frac{dV}{d\varphi}\delta\varphi$. The last contribution is the leading one during inflation. In this case it can be shown that δT^j_i to first order in $\delta\varphi$ is diagonal and the functions Φ and Ψ in (4.3) are equal to each other[34,35],

$$ds^2 = (1+2\Phi)d\tau^2 - (1-2\Phi)a^2(\tau)d\underline{x}^2 \tag{4.5}$$

The function $\Phi(x,\tau)$ is called relativistic potential; it is analogous to the Newtonian potential which is used for the description of weak gravitational fields. The coordinate system (4.5) is more convenient for the investigation of density perturbations than the usual synchronous system. After fixing the synchronous gauge by the condition $\delta g_{i0} = 0$, there still remains some freedom to change the coordinate system, which leads to the appearance of two fictitious perturbation modes. This leads to some ambiguity in the computation of density perturbations $\delta\rho$, which often lead to errors especially when density perturbations during the inflationary stage are considered. Fortunately, these modes give no contribution to the relativistic potential $\Phi(x,\tau)$, which is connected with the long-wave density perturbations ($k < H$) in the coordinate system (4.5) by the simple relation[34,35]

$$\frac{\delta\rho}{\rho} = -2\Phi\ . \tag{4.6}$$

In the first approximation in $\delta\varphi(x,\tau)$ and Φ, the mean field φ_c (4.4) satisfies the equations $\ddot{\varphi}_c + 3H\dot{\varphi}_c = -\frac{dV}{d\varphi}$ and $H^2 + \frac{k}{a^2} =$

$\frac{8\pi}{3M_P^2}\left(\frac{1}{2}\dot{\varphi}_c^2 + V(\varphi_c)\right)$, whereas for the fluctuations $\delta\varphi$ and Φ one can obtain the following system of equations[34]

(4.7)
$$\ddot{\Phi} + \left(\frac{\dot{a}}{a} - 2\frac{\ddot{\varphi}_c}{\dot{\varphi}_c}\right)\dot{\Phi} - \frac{1}{a^2}\Delta\Phi + 2\left(\frac{\ddot{a}}{a} - \left(\frac{\dot{a}}{a}\right)^2 - \frac{\dot{a}\dot{\varphi}}{a\varphi}\right)\Phi = 0\,,$$

(4.8)
$$\frac{1}{a}(a\Phi)_{,i} = \frac{4\pi}{M_P^2}(\dot{\varphi}_0\delta\varphi)_{,i}\,,$$

(4.9)
$$\delta\ddot{\varphi} + 3\frac{\dot{a}}{a}\delta\dot{\varphi} - \frac{1}{a^2}\Delta\delta\varphi + V''(\varphi_c)\delta\varphi + 2V'(\varphi_c)\Phi - 4\dot{\varphi}_0\dot{\Phi} = 0\,.$$

After some algebraic manipulations, these equations lead to the following equations for the long-wave density perturbations generated in the chaotic inflation scenario[34]

(4.10)
$$\frac{\delta\rho(\varphi_c)}{\rho} = \frac{V'(\varphi_c)}{V(\varphi_c)}\delta\varphi = -2\Phi$$

during inflation, whereas for the cold matter dominated universe after inflation one obtains the usual result[14] (up to some numerical coefficient[34]), which in the chaotic inflation scenario can be represented in the following useful form[36,37]

(4.11)
$$\frac{\delta\rho(\varphi_c)}{\rho} = \frac{48}{5}\sqrt{\frac{2\pi}{3}}\frac{V^{3/2}(\varphi_c)}{M_P^3 V'(\varphi_c)}$$

Here $\delta\rho(\varphi_c)$ is the amplitude of the perturbations which were generated (*i.e.* which acquire wavelength $l \gtrsim H^{-1}$ and become frozen in) when the field φ was equal to φ_c. For example, in the theory $V(\varphi) = \frac{\lambda}{4}\varphi^4$

(4.12)
$$\frac{\delta\rho(\varphi_c)}{\rho} \sim 2\sqrt{\lambda}\left(\frac{\varphi_c}{M_P}\right)^3\,.$$

The structure of the observable part of the universe was formed at values of φ_c, for which $\exp\left(\frac{\pi\varphi_c^2}{M_P^2}\right) \lesssim 65$, *i.e.* $\varphi_c \lesssim 4.5M_P$.

This gives

$$\frac{\delta\rho}{\rho} \lesssim 10^2\sqrt{\lambda} \tag{4.13}$$

in the observable part of the universe, which yields the constraint

$$\lambda \sim 10^{-14} \tag{4.14}$$

in order to have $\frac{\delta\rho}{\rho} \sim 10^{-5}$. Note that in this scenario there is *no* absolute upper bound on $V(\varphi)$, such as $V(\varphi) \lesssim 10^{-12}M_P^4$. The latter constraint[38,39] applies to the last stages of inflation only[37] (*i.e.* at $\varphi \lesssim 4.5M_P$), and it is automatically satisfied for $\lambda \sim 10^{-14}$.

From Eq. (4.12) it follows that after inflation $\frac{\delta\rho}{\rho} \gtrsim 0(1)$ for perturbations generated when $\varphi > \varphi^* \sim \lambda^{-1/6}M_P$. This is in agreement with our estimate (2.6), which shows that when $\varphi \gtrsim \varphi^*$ the role of quantum fluctuations of the field φ becomes especially important and self-reproduction of the universe becomes possible. However, one should emphasize, that *during inflation* the corresponding long-wave density perturbations $\frac{\delta\rho}{\rho} = -2\Phi$ (4.10) are of the order $\Phi \sim \frac{\delta\varphi}{\varphi} \ll 1$, and therefore one can in fact treat Φ as a small perturbation in Eqs. (4.7)-(4.9) even for $\varphi \gg \varphi^*$. This is one more advantage of the use of the coordinates (4.5) since using fictitious modes in certain versions of synchronous gauge formalism one could come to the wrong conclusion that if $\frac{\delta\rho}{\rho}$ after inflation is large (4.11), then the corresponding perturbations during inflation are also large.

To investigate the distribution $P_c(\varphi, \tau)$ one could use Eq. (4.9). During inflation the last term in this equation can be omitted, whereas the term $2V'\Phi$, according to (4.10), is equal to $-\frac{(V')^2}{V}\delta\varphi$. This additional term (which is absent in Eq.

(2.2)) can be omitted in the new inflationary universe scenario where during inflation V' is very small, but in the chaotic inflation scenario this term is of the same order as the term $V''\delta\varphi$. This makes an investigation of $P_c(\varphi, \tau)$ slightly more complicated than that performed in Sec. 3. However, for our purposes it is sufficient to note, that during the first stage of diffusion, for which the effects connected with the curvature of $V(\varphi)$ are negligibly small, all results obtained in the previous section remain valid also in the coordinate system (4.5). This is sufficient to confirm our statements concerning the regime of self-reproduction of the inflationary universe.

In order to confirm our conclusions by a different method, let us investigate the evolution of the field Φ by the methods used in the previous section. By means of Eqs. (4.6) one can easily show that during inflation Eq. (4.7) takes the form

$$\ddot{\Phi} + H(\varphi_c)\dot{\Phi} - \frac{1}{a^2}\Delta\Phi + M^2(\varphi_c)\Phi = 0 , \tag{4.15}$$

where

$$M^2(\varphi_c) = \frac{V'}{3}\left(-\frac{V'}{V} + \frac{2}{\varphi}\right) = -\frac{2\lambda\varphi^{2n-2}(n-1)}{3M_P^{2n-4}} \tag{4.16}$$

for the theories $V(\varphi) = \frac{\lambda\varphi^{2n}}{2nM_P^{2n-4}}$. Equation (4.15) is very similar to Eq. (2.2), which means that long-wave perturbations of the field Φ are generated during inflation. By comparing Eqs. (2.3) and (4.10) one can easily show that during the first stage of diffusion of the field Φ its dispersion grows as follows

$$\Delta_\Phi^2 = \langle \Phi^2 \rangle = \frac{\sqrt{2}(V')^2}{3\sqrt{3\pi V}M_P^3}\tau . \tag{4.17}$$

This, together with Eq. (4.15), leads to the diffusion equation

$$\frac{\partial P_c(\Phi, \tau)}{\partial \tau} = \frac{(V')^2}{3\sqrt{6\pi V}M_P^3}\frac{\partial^2 P}{\partial \Phi^2} + \frac{M^2(\varphi_c)}{H(\varphi_c)}\frac{\partial}{\partial \Phi}(P_c\Phi) . \tag{4.18}$$

This equation can be solved by the same methods as before. Here we will restrict ourselves to the investigation of the first stage, when the field φ_c decreases by $\Delta\varphi = \frac{\varphi_0}{N} \ll \varphi_0$. In this period Eq. (4.17) is valid, and

$$(4.19) \qquad P_c(\Phi, \tau) \sim \exp\left(-\frac{3\Phi^2\sqrt{3\pi V(\varphi_c)M_P^3}}{2\sqrt{2}\left(V'(\varphi_c)\right)^2 t}\right),$$

where, according to (2.2),

$$(4.20) \qquad \varphi_c = \varphi_0\left(1 - \frac{1}{2}\sqrt{\frac{n\lambda}{3\pi}\frac{\varphi_0^{n-2}}{M_P^{n-3}}}\tau\right).$$

At the end of this stage $\tau = \Delta\tau$, where

$$(4.21) \qquad \Delta\tau = \frac{2}{N}\sqrt{\frac{3\pi}{n\lambda}\frac{M_P^{n-3}}{\varphi_0^{n-2}}},$$

which gives

$$(4.22) \qquad P_c(\Phi, \Delta\tau) \sim \exp\left(-\frac{3N\Phi^2 M_P^{2n}}{8\pi\lambda\varphi_0^{2n}}\right) \sim \exp\left(-\frac{3N\Phi^2 M_P^4}{16n\pi V(\varphi_0)}\right).$$

During this time the mean value of $V(\varphi)$ decreases by $\Delta V = V(\varphi_0) \cdot \frac{n}{N} \ll V(\varphi_0)$ (for $N \gg 2n$). This means that the part of the initial (coordinate) volume remaining in the state with $V(\varphi) = V(\varphi_0)$ is given by Eq. (4.20) with $\Phi = -\frac{\Delta V}{2V} = -\frac{n}{N}$:

$$(4.23) \qquad P_c\left(-\frac{n}{N}, \Delta\tau\right) \sim \exp\left(-\frac{3nM_P^4}{16N\pi V(\varphi_0)}\right),$$

and the total volume of the universe remaining in the state with the same value of φ is proportional to

$$(4.24) \qquad P_p(\Phi, \Delta\tau) \sim P_c\left(-\frac{n}{N}, \Delta\tau\right) \cdot e^{3H\Delta\tau} \sim \exp\left(-\frac{3n^2 M_P^{2n}}{8N\pi\lambda\varphi_0^{2n}} + \frac{12\pi\varphi_0^2}{NnM_P^2}\right)$$

From (4.24) it follows again that the total volume filled with the field

$$\varphi \gtrsim \varphi^* = \lambda^{-1/(2n+2)} M_P$$

grows as the universe expands during the time $\Delta\tau$ (4.21). The same result remains valid after the next time interval $\Delta\tau$ etc. Thus, we have confirmed by this method our previous results concerning self-reproduction of inflationary domains of the universe with the field $\varphi > \varphi^*$.

However, in this approach we have not yet confirmed the statement contained in Secs. 2 and 3, that the major part of the physical volume of the inflationary universe after some sufficiently large time t of expansion should be in a state with a maximal possible value of φ (*i.e.* with $\varphi \sim \varphi_p$, for which $V(\varphi) \sim M_P^4$). The reason is that this statement is correct if we consider the evolution of the universe not with respect to the coordinate time τ (4.5), but with respect to the proper time t in a synchronous coordinate system of the type of (4.1) and (4.2).

To verify it, let us make a transformation to synchronous coordinates (t, x) which coincide with the coordinates (τ, x) at $t = 0$

$$(4.25) \qquad t = \tau + \int_0^\tau \Phi d\tau \ , \qquad x^\alpha = x^\alpha + \int_0^\tau \frac{d\tau}{a(\tau)} \int_0^\tau \Phi_{,\alpha} d\tau \ .$$

In a general case, the metric in the coordinates (4.25) is rather complicated. However, the main contribution to Φ is given by perturbations with exponentially large wavelengths. As a result, it can be shown that in the long-wave limit ($l \gg H^{-1}$) the corresponding metric is just given by Eqs. (4.1) and (4.2).

Thus, we have justified the implicit assumption which was made in Secs. 2 and 3, that during a sufficiently long time $(\Delta t \sim \Delta\tau\,(4.21))$ one can describe an inflationary domain of initial size $0(H^{-1})^*$ by the coordinate system (4.1). This is sufficient to confirm the main results obtained in these sections.

An alternative way to confirm the results obtained in Sec. 3 is to note that in these coordinates the proper (physical) volume grows approximately as $\exp(3H(\varphi)t) \sim \exp(3H(\varphi_c)(1-\Phi)t)$. At $\Phi \ll 1$ the distribution $P_c(\Phi, \Delta t)$ will approximately coincide with the distribution $P_c(\Phi, \Delta\tau)$ (4.22), whereas the distribution $P_p(\Phi, \Delta t)$ will be given by

(4.26)

$$P_p(\Phi, \Delta t) \sim P_c(\Phi, \Delta t) \exp(3H(\varphi_c)(1-\Phi)\Delta t)$$

$$\sim \exp\left(-\frac{3N\Phi^2 M_P^2}{16n\pi V(\varphi_0)}(1+0(\Phi)) + \frac{12\pi\varphi_0^2}{NnM_P^2}(1-\Phi)\right) ,$$

which implies that for $\varphi \gg \varphi^*$ the distribution $P_c(\Phi, \Delta t)$ grows with the growth of $-\Phi$, *i.e.* with the growth of φ. Actually, this equation is valid only for $\Phi \ll 1$ and for intervals Δt (4.21) with $\varphi_0 - \varphi_c(t) \ll \varphi_0$, but this is quite sufficient for our purposes, since from Eq. (4.26) it follows that for *each* time interval Δt the field φ averaged over the physical volume $\exp(3H(\varphi)t)$ grows, if initially $\varphi > \varphi^*$.

Thus, we see that the statement concerning the self-reproductio of inflationary mini-universes with $\varphi \gtrsim \varphi^*$ is gauge-independent, whereas the validity of the statement that the major part of the physical volume of the universe at a given time t should be in a state with a maximal possible field φ, depends on what is called time t. The physical meaning of our results obtained earlier in the synchronous coordinate system can be explained as follows.

---

* Strictly speaking, this is correct only when wavelengths of the perturbations $\delta\varphi$ and Φ become several times larger than H^{-1}, so that the contribution of the gradient terms $\sim \partial_\mu(\delta\varphi)\partial^\mu(\delta\varphi)$ to δT_μ^ν becomes smaller than $V'\delta\varphi$. This does not alter our main results.

Consider many comoving observers in an inflationary domain of the universe of initial size $l \sim 0(H^{-1})$, which contained a sufficiently big field φ and which was sufficiently homogeneous $\left(\varphi_c(0) = \varphi_0, \delta\varphi \sim H, \Phi \sim -\frac{V'}{2V}\delta\varphi \ll 1\right)$ at the moment $t = 0$, when clocks of all observers were synchronized. The total number of observers which see themselves in a state with $\varphi > \varphi_0$ at some moment t by their clocks is proportional to $P_c(\varphi, t)$, which decreases at large φ. At that time the distance between the observes grows as $\exp\left(H(\varphi)t\right)$, and the total volume occupied by the observers in a state with the field φ is proportional to $P_p(\varphi, t)$, which after the time Δt (4.21) for $\varphi > \varphi^*$ starts to grow as $\exp\left(C\frac{\varphi^2}{M_P^2}\right), C = 0(1)$.

One can interpret our results in yet another way. If the universe is self-reproducing, then the standard question concerning *initial conditions in the whole universe* may become irrelevant, since in such a universe there may be no global initial spacelike singular hypersurface at all [18–21]. We cannot say as before that our universe was created *as a whole* about 10^{10} years ago, and that we are interested in the initial conditions at that time. However, we can say, that the density of matter in any particular domain of the universe, which sometimes was inflationary, becomes equal to the energy density in the observable part of our universe $\rho_0 \sim 10^{-29}\ g \cdot cm^{-3}$ about 10^{10} years *after the end of inflation in this domain.* Just at the time of that order, life of our type becomes possible and intelligent observers of our type first appear. The total number of all intelligent observers *of our type* is, presumably, proportional to the total volume of all domains of the universe with density $\rho \sim 10^{-29}\ g \cdot cm^{-3}$. Each such observer at that time can see the part of the universe of the size $ct \sim 10^{28}$ cm. Then, each such observer (including ourselves) may ask the question of the history of his own domain (of the present size $\sim 10^{28}$ cm) during inflation, and the probability that a domain of a size $\sim 10^{28}$ cm with density $\rho \sim 10^{-29}\ g \cdot cm^{-3}$ has originated from some inflationary domain containing the field $\varphi = \varphi_0$.

If the corresponding inflationary domain was created from some (local) initial singularity or from space-time foam, then as follows from Eq. (3.32), a typical initial value of the field φ in this domain was of the order of $\varphi \sim \varphi_p$, where $V(\varphi_p) \sim M_P^4$. However, there also exists the possibility that this domain was created from the previously existing inflationary domains by the process of the self-reproducing universe. This means that the value of the field φ in this domain initially should be greater than φ^*. Now we are going to argue that most domains with density $\rho \sim 10^{-29}$ g cm^{-1} and size $0(10^{28}$ cm) were created from domains in which the field $\varphi > \varphi^*$ at least once was as large as φ_p.

The reason is that during the time Δt (4.21) the probability that the field φ_0 jumps to $\varphi \gtrsim \varphi_0$ when $\varphi_c(t) \approx \varphi_0$) is suppressed by the factor

$$P_c(\varphi, \Delta t) \sim \exp\left(-\frac{cM_P^4}{V(\varphi_0)}\right) , \tag{4.27}$$

where $c = 0(1)$ (see Eq. (4.23)). (For $t \gg \Delta t$ the probability of such a jump becomes strongly suppressed since $\varphi_c(t)$ becomes much smaller than φ_0.) After inflation the physical volume of those domains, in which the field φ did not jump, grows $\exp\left(\frac{6\pi\varphi_0^2}{nM_P^2}\right)$ times $\left(\text{see Eq. (2.5)}\right)$, whereas the physical volume of those domains, in which the field jumped to $\varphi > \varphi_0$, grows $\exp\left(\frac{6\pi\varphi^2}{nM_P^2}\right)$ times. This implies that most domains of a given size (*e.g.* $0(10^{28}$ cm)) and of a given density (*e.g.* $10^{-29}\ g \cdot cm^{-3}$), which are formed in the process of self-reproduction of the inflationary domain with $\varphi > \varphi^*$, originate from domains with maximal possible value of the field φ, *i.e.* with $\varphi \sim \varphi_p$. In other words, even if initially the field φ in an inflationary domain was smaller than φ_p (but bigger than φ^*), then most domains with $\rho \sim 10^{-29}\ g \cdot cm^{-3}$ (*i.e.* most domains at the moment when intelligent observers of our type first appear) are formed as a result of inflation of those parts of the initial domain, in which the field φ jumped to $\varphi \sim \varphi_p$. This means in particular that it is most probable that the observable part of our universe has originated from a domain with $V(\varphi) \sim M_P^4$. One of the applications of these results will be considered in the subsequent section.

5. Chaotic Inflation with Constrained Fields

One of the most important problems of inflationary cosmology is the problem of initial conditions which are necessary for inflation to start. In the theory $\frac{\lambda}{4}\varphi^4$ with $\lambda \sim 10^{-14}$ (which is necessary in order to obtain a density perturbation $\frac{\delta\rho}{\rho} \sim 10^{-5}$ in the observable part of the universe) the initial condition $V(\varphi) \sim M_P^4$ implies that the initial amplitude of the field φ at the Planck time $t_P \sim M_P^{-1}$ should be of the order of $\lambda^{-\frac{1}{4}}M_P \sim 10^3 M_P$. This would mean that, in accordance with the condition $(\partial_i\varphi)^2 \leq M_P^4$, which should be satisfied in any domain of classical space-time, the field φ should remain as large as $10^3 M_P$ in a domain of a size $\ell \geq \lambda^{-\frac{1}{4}}M_P$, which is much bigger than the size of a causally connected region $\ell_P \sim M_P^{-1}$ at that time. It was argued that in order to avoid this "acausal correlation" between values of φ in different domains, the field φ initially should not be much bigger than M_P (see *e.g.* ref. 40).

The resolution of this complication was given in ref. 31 (see also Chapter 1 of this book). There is no reason to expect that the energy density ρ can simultaneously become smaller than M_P^4 in all causally disconnected parts of a domain much larger than M_P^{-1}, since *this* would imply an acausal correlation. Therefore, the typical initial size of a classical space-time must be of the order M_P^{-1}. Such domains look like separate islands of classical space-time rising out of the space-time foam. Outside each domain the gradient of φ is not restricted to be smaller than M_P^2, and there is no acausal correlation between the values of the field φ in different domains. A simple way to understand this situation is to consider a closed universe. Its typical initial size at the Planck time is $0(M_P^{-1})$, *i.e.* it consists of one or two causally connected regions, and no problem of acausal correlation appears in this case, independent of the initial value of the field φ inside this universe.

The absence of any *a priori* constraints on the initial value of the field φ can be most easily understood by considering a theory with $V(\varphi) = \text{const}$. This theory is invariant under the shift $\varphi \to \varphi + c$, which means that *any* value of the field φ can appear in the early universe with the same probability. Thus, the only constraint on the amplitude of the field φ comes from the inequality $V(\varphi) \leq M_P^4$, which gives the only natural scale for the initial value of the field $\varphi : V(\varphi_0) \sim M_P^4$.

One must take into account, however, that different people sometimes have different notions of what is natural and what is not. Therefore in the present section we are going to show that the chaotic inflation scenario can be easily realized in realistic models of elementary particles even if there were constraints on the initial values of the scalar fields φ of the type of $\varphi \leq CM_P$. The exact value of the constant C depends on the choice of the effective potential of the scalar fields driving inflation, but typically $C = 0(5)$. Another motivation to investigate inflation with constrained scalar fields is that there may exist some reasons which would preclude the existence of scalar fields with $\varphi \geq CM_P$ in our universe. For example, one might expect that quantum gravity effects could lead to corrections to $V(\varphi)$ of the type of $\sum_n C_n(\varphi/M_P)^n$, which would make $V(\varphi)$ much bigger than M_P^4 for $\varphi \geq 0(M_P)$ [41]. Fortunately this is not the case: these corrections are of the type of $C_1 M_P^{-2} \frac{d^2V}{d\varphi^2} \cdot V(\varphi) +$

$C_2 M_P^{-4} V^2(\varphi) \ll V(\varphi)$ for $\lambda \ll 1, V(\varphi) \ll M_P^4$ [42]. The real problem may appear in theories in which the effective potential for some reasons includes factors of the type $\exp(\varphi^2/CM_P^2)$, which precludes inflation with $V(\varphi) \sim M_P^4$ in theories with $\lambda \ll C^{-2}$. Another complication may appear in the theory of a scalar field φ with the Lagrangian $\mathcal{L}(\varphi) = RM_P^2/16\pi + L(\varphi) - \xi/2R\varphi^2$. Inflation is quite possible in this theory for $\xi < 0$ [43], but for $\xi > 0$, one may encounter some problems[44], since at $\varphi = M_P/\sqrt{8\pi\xi}$ the effective gravitational constant G has a pole[45], which leads to a cosmological singularity preventing motion of the field φ from the region $\varphi > M_P/\sqrt{8\pi\xi}$ to $\varphi = 0$ [46]. It would be interesting therefore to investigate whether it is possible to realize the chaotic inflation scenario in a natural way if there exist some constraints on the values of the fields φ of the type $\varphi \leq CM_P$. In what follows we are going to give a simple argument (see also refs. 31,47) which shows that this is actually possible when the process of self-reproduction of the universe in the chaotic inflation scenario is taken into account.

We will consider as an example a simple theory of two fields φ and Φ with effective potentials which for $\Phi, \varphi \leq CM_P$ are given by $\lambda_\Phi \Phi^4$ and $\lambda_\varphi \varphi^4$ respectively. The field φ plays the role of the standard inflaton field with $\lambda_\varphi \sim 10^{-14}$, which should drive the last stages of inflation, whereas the field Φ may be just the ordinary Higgs field in the theory of electroweak interactions, with $\lambda_\Phi \sim 10^{-2}$.

If there were no field Φ in the theory, the constraint $\varphi \leq CM_P$ would make inflation unnatural, since at the Planck time, when the total density of matter is of the order of M_P^4, the value of $V(p)$ would be much smaller. In such a case a typical closed universe would collapse before the beginning of inflation. In the presence of the field Φ, with $\lambda_\Phi \sim 10^{-2}$, inflation may naturally start with $V(\varphi) \sim M_P^4$ even if the field Φ is constrained to be smaller than CM_P with $C \sim 1$. In that case the first stage of expansion will be driven by the field Φ, and there will be no problem with the *beginning* of inflation.

In order to investigate the subsequent behavior of the inflationary universe in which $V(\Phi)$ initially was (almost) as large as M_P^4 one should make a clear distinction between two different questions to be asked. The first question concerns the typical behavior of the fields φ and Φ at a given point x as a function of time t. The second question is: Which part of the physical volume of the universe appears as a result of inflation of domains containing given fields φ and Φ? The answer to the first question is very simple. For rather typical initial conditions in the very early universe, the time of rolling of the field Φ down to the minimum of $V(\Phi)$ (*i.e.* the average time of inflation driven by the field Φ *at the given point* x) is $0(\sqrt{6\pi}/\sqrt{\lambda_\Phi}M_P^{-1})$ (see eq. (2.8) of Chapter 2). This answer remains true after taking into account the quantum fluctuations of the fields Φ and φ discussed in previous sections.

The answer to the second question is much more complicated, and here quantum fluctuations of scalar fields Φ and φ play a very important role. As it was shown in the previous Section, the main part of the physical volume of the universe appears due to exponential expansion of those inflationary domains of the universe, in which the field Φ was permanently fluctuating at a value near $\Phi \sim \lambda_\Phi^{-1/4} M_P$ (*i.e.* at $V(\Phi) \sim M_P^4$), see Sec. 4. The Hubble constant H in these domains of the universe remains of the order M_P for a very long (or infinitely long) time. As discussed in the previous sections, in such domains long-wave fluctuations of the field φ are produced with the dispersion

$$\bar{\varphi} = \sqrt{<\varphi^2>} \sim \frac{H^2}{m_\varphi} \tag{5.4}$$

where $m_\varphi = \sqrt{3\lambda_\varphi}\varphi$ is the effective mass of the scalar field φ.

This gives

(5.5) $$\overline{\varphi} \sim \lambda_{\varphi}^{-1/4} M_P \ .$$

In other words, the major part of the volume of the universe appears as a result of inflation of domains in which field Φ for a long time fluctuates near the value $\Phi \sim \lambda_{\Phi}^{-1/4} M_P$, and the field φ acquires the value $\varphi \sim \lambda_{\varphi}^{-1/4} M_P$. If the field φ for some reason is constrained to be in the region $0 < \varphi \leq C M_P$, then this field can acquire any value in this region with approximately the same probability. However, the main contribution to the final volume of the universe comes from the domains with the maximal possible values of the field φ. Indeed *at any given point* x the heavy field Φ rolls down to the minimum of $V(\Phi)$ at $\Phi = 0$ much more rapidly than the light field φ. Therefore though the field Φ drives the first stages of inflation, the last stages of inflation typically are driven by the field φ. The growth of the volume of the universe during this stage of inflation is proportional to $\exp(3\pi\varphi^2/M_P^2)$ (see Eq. (2.9) of Chapter 2), which means that the main contribution to the volume of our universe comes from the domains with $\varphi \sim C M_P$. As it is noted in Capter 2, for $\varphi \geq 5 M_P$, the size of any domain of initial size $\ell \geq M_P^{-1}$ after inflation becomes bigger than the size of the observable part of the universe. Therefore the density perturbations, which are generated in this scenario for $C \geq 5$, in the observable part of the universe will be determined by fluctuations of the field φ and will have a desirable amplitude $\delta\rho/\rho \sim 10^{-5}$ for $\lambda_{\varphi} \sim 10^{-14}$.

This means that in a realistic theory, in which there exist several types of scalar fields φ_i some of which are sufficiently heavy or have sufficiently large coupling constants, initial conditions for chaotic inflation are natural even if for some reasons these fields (initially) should be rather small, $\varphi_i \leq C M_P$. In such a situation the first stages of inflation typically are driven by most heavy fields φ_i, whereas the last stage of inflation typically is driven by the lightest of the scalar fields φ_i, which determines the large-scale structure of the observable part of the universe.

6. The Universe with Fluctuating Dimension, Baby Universe Formation and Stochastic Approach to Inflation

It is important that the process of self-reproduction of the inflationary mini-universes can occur at densities much smaller than the Planck density. For example, in the theory $V(\varphi) = \frac{\lambda\varphi^4}{4}$ this process occurs for $V(\varphi) > V(\varphi^*) \sim \lambda^{1/3}M_P^4 \sim 10^{-5}M_P^4$ (for $\lambda \sim 10^{-14}$). Therefore $V(\varphi^*) \ll M_P^4$, which means that in order to prove the very existence of the regime of self-reproduction there is no need to make any hypotheses about physical processes at $V(\varphi) \gtrsim M_P^4$.

On the other hand, according to our results, a considerable part of the physical volume of the self-reproducing inflationary universe even now should be in a state with $V(\varphi) \sim M_P^4$. In such domains (or in domains where $V(\varphi)$ is somewhat smaller than M_P^4, so that the above-mentioned results are reliable) fluctuations of all other types of scalar fields Φ_i and of the metric on a scale $H^{-1} \sim M_P^{-1}$ become very large. These fluctuations may transfer Φ_i from one local minimum of $V(\varphi, \Phi_i)$ to another, thus creating inflationary mini-universes with all possible types of symmetry breaking between fundamental interactions[49]. Moreover, large fluctuations of metric may lead to a *local* size $0(H^{-1})$. If these domains remain inflationary after this process, the universe becomes divided into many exponentially large domains (mini-universes) with all possible types of compactification[17–21,50].

According to this scenario, we live in the mini-universe with four-dimensional space-time and with the low-energy particle physics of our type not because different mini-universes are impossible, but because life of our type cannot exist in domains with different dimensionality and different laws of particle physics.

In order to explain why we do not see other domains of the universe, let us note that all parts of the hot universe with $\frac{\delta\rho}{\rho} \lesssim 1$ are formed only from those inflationary domains, in which the field φ becomes smaller than φ^*. At that moment the size of any inflationary domain is $l \gtrsim H^{-1}(\varphi^*) \gg M_P^{-1}$. According to Eq. (2.6), after inflation the size of each mini-universe in the theory $V(\varphi) = \frac{\lambda\varphi^4}{4}$ exceeds

(6.1)

$$l \gtrsim M_P^{-1} \exp\left(\pi \left(\frac{\varphi^*}{M_P}\right)^2\right) \sim 10^{-33}\text{cm}\cdot\exp(\pi\lambda^{-1/3}) \gtrsim 10^{3\cdot 10^6}\text{cm}$$

for $\lambda \sim 10^{-14}$. This is much greater than the size of the observable part of our universe $l_0 \sim 10^{28}$ cm. That is why the properties of space-time and of elementary particles are the same in all parts of the universe which we see now. However, on a much bigger scale the universe is very inhomgeneous, and the laws of low-energy physics and even the dimensionality of space-time may be different in different exponentially large domains of the universe.

This idea is rather general. However, we shall give a concrete example which shows that inflation actually may lead to creation of exponentially large domains of different dimension. Let us consider $d = 6$ Einstein-Maxwell-scalar theory, which admits a compactification $M_4 \times S_2$, where M_4 is $d = 4$ Minkowski space-time and S_2 is a two-sphere of a radius b_0. The corresponding action can be represented in the following form:

$$S = N \int d^6x\sqrt{-g_6}\Big\{\delta R - \Lambda - \frac{1}{2}\partial_M\varphi\partial^M\varphi - V(\varphi) - \frac{1}{4}F_{MN}F^{MN}\Big\} \,. \tag{6.2}$$

Here $F_{MN} = \partial_M A_N - \partial_N A_M, \delta^M_M = 6, V(0) = 0, \partial V/\partial\varphi|_{\varphi=0} = 0, \delta = M_P^2/(16\pi)$, N is the normalization constant, and Λ is the cosmological constant (to be determined below).

In order to describe compactification of $d = 6$ space-time down to $d = 4$ we will consider space-time with metric, which (locally) can be represented as follows:

$$dS^2 = -dt^2 + a^2(t)g_{ij}(x^k)dx^i dx^j + b^2(t)g_{ab}(x^c)dx^a dx^b \,, \tag{6.3}$$

where $i, j, k = 1,2,3; a, b, c = 4,5,$

$$\begin{aligned} g_{ij}(x^k)dx^i dx^j &= (dx^1)^2 + (dx^2)^2 + (dx^3)^2, \\ g_{ab}(x^c)dx^a dx^b &= d\theta^2 + \sin^2\theta d\alpha^2 \,. \end{aligned} \tag{6.4}$$

One of the solutions of the equation for the vector field is the so-called "monopole" configuration of A_M [51]. All components of this field are equal to zero except $A_\alpha = f(1 - \cos\theta)$, where f is some constant . The nonvanishing components of F_{MN} are given by

$$F_{ab} = f\sqrt{\det g_{ab}}\;\epsilon_{ab} \,, \tag{6.5}$$

where ϵ_{ab} is the Levi-Civita tensor. If the field A_M interacts with some charged particles, the parameter f becomes quantized, just as a magnetic charge of monopole. However, in the present paper we will not identify A_M with the ordinary electromagnetic field and will consider f as a free parameter.

Let us assume that φ is x-independent, like a and b. The Einstein equations look as follows:

$$3\frac{\ddot{a}}{a} + 2\frac{\ddot{b}}{b} = \delta^{-1}\left\{\frac{1}{4}(\Lambda + V) - \frac{1}{2}\dot{\varphi}^2 - \frac{1}{8}f^2 b^{-4}\right\}, \tag{6.6}$$

$$\frac{\ddot{a}}{a} + 2\frac{\dot{a}^2}{a^2} + 2\frac{\dot{a}}{a}\frac{\dot{b}}{b} = \delta^{-1}\left\{\frac{1}{4}(\Lambda + V) - \frac{1}{8}f^2 b^{-4}\right\}, \tag{6.7}$$

$$\frac{\ddot{b}}{b} + \frac{\dot{b}^2}{b^2} + 3\frac{\dot{a}}{a}\frac{\dot{b}}{b} = \delta^{-1}\left\{\frac{1}{4}(\Lambda + V) + \frac{3}{8}f^2 b^{-4} - b^{-2}\right\}, \tag{6.8}$$

$$\ddot{\varphi} + \dot{\varphi}\left(3\frac{\dot{a}}{a} + 2\frac{\dot{b}}{b}\right) + \frac{\partial V}{\partial \varphi} = 0\,. \tag{6.9}$$

The ground state of the type $M_4 \times S_2$ corresponding to the minimum of $V(\varphi)$ at $V(\varphi) = 0$ exists if $\Lambda = 2\delta^2 f^{-2}$. In this case the radius of compactification is equal to

$$b_0 = f(2\delta)^{-1/2} = (\delta/\Lambda)^{1/2}\,. \tag{6.10}$$

(One can check that in order to have a standard normalization of the effective four-dimensional action one should take $N = (4\pi b_0^2)^{-1}$ in Eq. (6.2)).

Quantum gravity corrections to Eqs. (6.6)-(6.9) are small if $b_0 \gtrsim M_p^{-1}$ and if $V(\varphi) \lesssim M_p^4$. After a simple rearrangement, Eq. (6.8) becomes like the equation of motion of a particle with

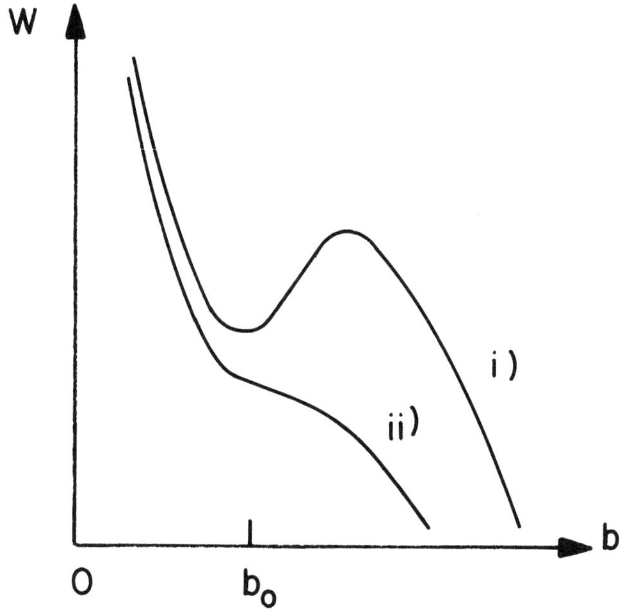

FIG. 3.1: Sketch of the effective potential $W(b)$ for $V(\varphi) = 0$ (i)) and $V(\varphi) > \frac{\Lambda}{3}(ii)$).

potential energy W and nonlinear friction,

$$\ddot{b} + \dot{b}\left(3\frac{\dot{a}}{a} + \frac{\dot{b}}{b}\right) + \frac{\partial W(b,\varphi)}{\partial b} = 0 \,, \tag{6.11}$$

$$\begin{aligned} W(b,\varphi) &= \frac{3}{8}\frac{b_0^2}{b^2} + \ln\frac{b}{b_0} \\ &\quad - \frac{1}{8}\left(1 + \frac{V(\varphi)}{\Lambda}\right)\frac{b^2}{b_0^2} \,. \end{aligned} \tag{6.12}$$

The shape of W for different values of $V(\varphi)$ is shown in Fig. 3.1. Let us now consider solutions of Eqs. (6.6)-(6.9) for a slowly changing field φ $\left(\left|\frac{\dot{\varphi}}{\varphi}\right| \ll \frac{\dot{a}}{a}\right)$ in the theory $V(\varphi) = \frac{1}{4}\lambda\varphi^4$. One can easily check that any such solution rapidly approaches one of the two asymptotics described by the following equations:

$$
\begin{aligned}
I. \quad & a \approx a_0 \exp\left\{\int^t H_I(\varphi)dt\right\} , \\
& b \approx b_0\, 3^{1/2}\left\{2 + \left(1 - 3\frac{V(\varphi)}{\Lambda}\right)^{1/2}\right\}^{-1/2} , \\
& 3H_I(\varphi)\dot{\varphi} \approx -\frac{\partial V(\varphi)}{\partial\varphi} , \\
& H_I(\varphi) = b_0^{-1}3^{-3/2}\left\{1 + 3\frac{V(\varphi)}{\Lambda} - \left(1 - 3\frac{V(\varphi)}{\Lambda}\right)^{1/2}\right\}^{1/2} ,
\end{aligned} \tag{6.13}
$$

for $16\delta^2\lambda = V_1 \ll V(\varphi) < \Lambda/3 = \frac{2}{3}\delta^2 f^{-2}$.

$$
\begin{aligned}
II. \quad & a \approx a_1 \exp\left\{\int^t H_{II}(\varphi)dt\right\} , \\
& b \approx b_1 \exp\left\{\int^t H_{II}(\varphi)dt\right\} , \\
& 5H_{II}(\varphi)\dot{\varphi} \approx -\frac{\partial V(\varphi)}{\partial\varphi} , \\
& H_{II}(\varphi) = b_0^{-1}20^{-1/2}\left(1 + \frac{V(\varphi)}{\Lambda}\right)^{1/2} .
\end{aligned} \tag{6.14}
$$

Here a_0 , a_1 , b_1 are some constants.

The first solution corresponds to the radius b of compactified space fixed at the local minimum of W. The effective dimension of space-time in this case is equal to four. The second solution describes b rolling down the right slope of W (see Fig. 3.1). For large V $\left(V(\varphi) > \frac{\Lambda}{3}\right)$ only this solution exists. It is important that though the initial values a_1 and b_1 may be quite different, all local properties of space-time (6.14) after a time interval $\tau \gg H_{II}^{-1}$ become almost undistinguishable from the properties of a six-dimensional de Sitter space-time dS_6.

At the classical level, the choice of a particular solution (6.13) or (6.14) corresponding to different dimensions of space-time (M_4 or dS_6) is uniquely determined by the choice of initial conditions in the very early stages of the evolution of the universe. In the context of the eternal chaotic inflation scenario the situation is quite different.

Any domain of the universe described by the solution (6.13) with $V(\varphi) \gtrsim V_2 \simeq 380\delta^2\lambda^{1/3}$ permanently reproduces itself and other inflationary domains with all possible values of $V(\varphi)$ according to our discussion in previous sections. However, in those domains in which the value of $V(\varphi)$ becomes bigger than $\frac{\Lambda}{3}$, the effective potential $W(b, \varphi)$ has no minimum at finite b, the process of decompactification starts, and very soon these domains acquire the same (local) structure as a six-dimensional inflationary universe (6.14).

As a result, the universe permanently splits into many exponentially large domains, in some of which space-time is four-dimensional, whereas in other domains space-time is six-dimensional. Note, that this process occurs in our model only if $V_2 < \frac{\Lambda}{3}$, since otherwise the four-dimensional universe cannot be self-reproducing. In the theory $\frac{1}{4}\lambda\varphi^4$ this implies that $b_0 \lesssim 3.10^{-2}\lambda^{-1/6}\delta^{-1/2}$. In order to obtain the necessary amplitude of primordial density fluctuations after inflation, $\frac{\delta\rho}{\rho} \sim 10^{-5}$, we must take $\lambda \sim 10^{-14}$, which means that the radius of compactification b_0 should not be bigger than $0(10)M_P^{-1}$.

One of the implications of the results obtained above is that the global structure of the universe cannot be understood in the context of the minisuperspace approach to quantum cosmology. From our investigation it follows, however, that some of the wave functions suggested in the minisuperspace approach can be useful for the investigation of sufficiently homogeneous parts of the universe. For example, the Hartle-Hawking wave function[3] is useful for the investigation of diffusion (=tunneling) from metastable vacuum states φ_i with $V(\varphi_i) > 0$, whereas the wave function describing quantum creation of the universe from "nothing"[5–8] is useful for investigating diffusion from the space-time foam with $V(\varphi_0) \sim M_p^4$, as well as for describing the first stage of diffusion from $\varphi = \varphi_0$ even if $V(\varphi_0) \ll M_p^4$.

The methods discussed here are rather general. However, we are still at the very beginning of the understanding of the global structure of the self-reproducing inflationary universe. For obvious reasons the universe consisting of inflationary domains producing new inflationary domains, producing new inflationary domains, etc., can be called *the fractal universe.* One may try to determine the Hausdorff dimension of space-time on the exponentially large length scale and to use new powerful methods developed in the related branches of mathematics[52].

One may wonder also whether methods analogous to those used here could be applied to investigate such hypothetical processes as the creation of an inflationary universe in a laboratory, or just in empty space.

The answer is still cautious. The universe can hardly be created in a laboratory at the classical level, *i.e.* without an account taken of quantum effects[53]. However, one may imagine that an inflationary domain of size $l \gtrsim 0(H^{-1})$ can be created due to quantum fluctuations in Minkowski space, and then it may become disconnected from the parent universe. The field φ inside this domain should be sufficiently homogeneous, $(\partial_i\varphi)^2 \lesssim V(\varphi)$, which means that its typical size l should be of the order $l \sim \varphi V^{-1/2}$. The field φ inside this domain should be quasistatic, which implies as usual that $\varphi \gtrsim M_P$ (in the theories $V(\varphi) \sim \varphi^n$). The remaining step is to calculate the dispersion Δ of quantum fluctuations with a wavelength $l \gtrsim \varphi V^{-1/2}$ in Minkowski space and the probability $P(\varphi)$ of creation of a domain of a size l filled with a sufficiently homogeneous classical field φ. One can easily verify that in Minkowski space $\Delta^2(l) \sim l^{-2} \sim \frac{V(\varphi)}{\varphi^2}$, which gives the following estimate for the probability of quantum creation of such a domain[60]

$$P(\varphi) \sim \exp\left(-\frac{\varphi^4}{V(\varphi)}\right) . \tag{6.15}$$

where $\varphi \gtrsim M_P$. In particular, in the theory $V(\varphi) = \frac{\lambda}{4}\varphi^4$ this equation reads

$$P(\varphi) \sim \exp\left(-\frac{1}{\lambda}\right) . \tag{6.16}$$

Note that this estimate differs considerably from the estimate of the probability of quantum creation of inflationary universe from "nothing," Eq. (3.32). The reason is that Eq. (6.2) describes the probability of diffusion from Minkowski space, whereas Eq. (3.32) describes diffusion from exponentially expanding (de Sitter) space. This difference implies also that Eq. (6.2) does not necessarily describe creation of an exponentially *expanding* domain of de Sitter space, *i.e.* a domain of inflationary universe (de Sitter space can be exponentially contracting as well). Therefore more work is needed in order to investigate whether quantum creation of an inflationary baby universe actually can occur in Minkowski space, though in our opinion this process is quite possible. (Another approach to the baby universe formation based on the use of euclidean methods was suggested in refs. 69-71, see Sec. 8.)

One can also speculate that the process of the inflationary baby universe production occurs in our universe right now in those its regions where quantum diffusion to the space-time foam with the Planck density takes place, since the diffusion backwards from the space-time foam produced in these regions might be interpreted as a formation of new inflationary universes.

If a consistent theory of inflationary universe production either from Minkowski space or from the space-time foam will be developed, it may become possible to speak about stationary (time-independent) solutions of diffusion equation, though these solutions will be different from Eq. (3.7). But in any case, as we have already seen, even the nonstationary solutions of diffusion equation lead to the stationary process of self-reproduction of inflationary mini-universes. In this sense we have already obtained a kind of stationarity, not in terms of time but in terms of the properties of the universe at a given density and on a given (physical) length scale. This kind of stationarity, as opposed to the stationarity in time of a ground state in the usual Minkowski space quantum field theory, may serve as a basis for the search of a correct wave function of the universe.

7. Life After Inflation

The history of cosmology knows many doomsday predictions: life in the universe will disappear after the stars stop shining, or after baryons decay, or after the universe collapses as a whole. However, after several billion years, when the Sun dims, we will probably find some alternative sources of energy. 10^{30} (or 10^{40}) years later, when baryons start to decay, there will be almost as many positrons as electrons. In principle, it is quite possible (and at that time it probably will not be technologically difficult) to convert positrons back to protons by making the positron density three orders of magnitude bigger than the nuclear density (in order to make their Fermi energy bigger than the proton mass) and using monopoles as a catalyzer of baryon formation *via* the inverse Rubakov effect. Finally, as we have discussed above, in most versions of the inflationary universe scenario the universe as a whole never collapses, but rather enters a regime of eternal reproduction of new inflationary domains, which later give rise to an infinite number of exponentially large regions with new stars and baryons, where life appears again and again.

Let us consider again a particular example of chaotic inflation in a theory with effective potential $V(\varphi) = \frac{m^2}{2}\varphi^2$. As it was shown in Chapter 1, the classical scalar field φ in this theory decreases very slowly from its initial value $\varphi_0 > M_P$,

$$\varphi(t) = \varphi_0 - \frac{mM_P}{2(3\pi)^{1/2}}t, \tag{7.1}$$

where M_P is the Planck mass, whereas the size of the universe grows as

$$a(t) = a_0 \exp\left(\frac{2\pi}{{M_P}^2}(\varphi_0^2 - \varphi^2(t))\right) . \tag{7.2}$$

This expansion is quasiexponential, which means that during time intervals a few times bigger than the typical time $\tau = H^{-1} = (3/\pi)^{1/2}\frac{M_P}{2m\varphi}$, where H is the Hubble parameter, the expansion of the universe can be described by the simple function $a(t) = a_0\exp(Ht)$. During the time H^{-1} the universe grows e times, and its volume grows $e^3 \sim 20$ times. At the same time the classical field $\varphi(t)$ decreases by $\Delta\varphi = M_P^2/2\pi\varphi$, but simultaneously inhomogeneities $\delta\varphi(x)$ are produced with a wavelength bigger than the size of the event horizon H^{-1} and with an average amplitude $|\delta\varphi| \sim H/2\pi \sim \frac{m\varphi}{(3\pi)^{1/2}M_P}$. Note, that $|\delta\varphi| \gg \Delta\varphi$ for $\varphi > \varphi^*$, where

$$\varphi^* = M_P \left(\frac{M_P}{m}\right)^{1/2} . \tag{7.3}$$

This means that during the time interval $\tau = H^{-1}$ the total volume of inflationary domains containing the field $\varphi \geq \varphi^*$ grows $0\left(\frac{e^3}{2}\right) \sim 10$ times. Therefore inflationary domains with $\varphi \geq \varphi^*$ unceasingly reproduce themselves. This means that there is no end to the evolution of the universe in this scenario, and that in the course of time new and newer exponentially large domains of the universe capable of supporting life will be produced.

These results imply that life in the inflationary universe will never disappear. Unfortunately, this conclusion does not automatically mean that one can be very optimistic about the future of mankind. Even though new domains of our type are permanently produced during inflation, later on each particular domain either will become practically empty and unsuitable for life (if the vacuum energy density is positive), or it will evolve into a huge black hole and collapse (if the vacuum energy density is negative, or if it vanishes) [59].

The physical meaning of the last statement can be explained as follows. Since an inflationary universe containing at least one sufficiently homogeneous domain with $\varphi > \varphi^*$ exists without end, its different parts go through the state with energy density $\rho = \rho_0 \sim 10^{-29} g \cdot cm^{-3}$ at different times (ρ_0 is the present energy density in the observable part of the universe). This means that we do not actually know what the total age of the part of the universe which we live now in is; we just know how much time ago the field φ became smaller than φ^* in our domain. Therefore instead of comparing the properties of the universe at a given time after the Big Bang, which may have no precise meaning in our model, one should compare the properties of different parts of the universe on a hypersurface with a given energy density ρ. As was shown in Sec. 4, the main part of the physical volume of the universe on the hypersurface with given ρ appears due to the expansion of those domains with $\varphi > \varphi^*$, where the field φ increases due to quantum fluctuations $\delta\varphi(x) > 0$, then rolls down and then increases again for as long a time as possible[54]. This implies that it is most probable that on a scale $l > l^*$, where l^* is the size of the part of the universe produced by inflation from a domain of size $H^{-1}(\varphi)$ filled with the field $\varphi \sim \varphi^*$, we live in a local *maximum* of energy density ρ, $\rho(x) = \rho_0 + \delta\rho(x), \delta\rho(x) > 0$. Some parts of the original inflationary domain of size H^{-1} with $\varphi > \varphi^*$ are even now in the inflationary phase, and in some other parts the energy density is as small as $\rho_o \sim 10^{-29} g \cdot cm^{-3}$. Thus, even though the density perturbations on the scale of the observable part of the universe are very small, $\delta\rho/\rho \sim 10^{-5}$ for $m \sim 10^{-6} - 10^{-7} M_P$, the density perturbations on a scale $l > l^*$ are very large[55], $\delta\rho/\rho \geq 1$. This conclusion, as well as the value of the length scale l^* on which the density perturbations become large (to be given below, eq.(7.4)), are in agreement with the results of computations of the magnitude of density perturbations in the inflationary universe performed in ref. 14.

In theories with vanishing vacuum energy density ρ_v, domains of the universe with $\delta\rho > 0$, $\delta\rho/\rho \geq 1$, with size $l(t)$ bigger than the size of the particle horizon, evolve into black holes at the moment when the size of the horizon $\sim t$ becomes bigger than $l(t)$ [56,57]. (Note that $l(t)$ grows as the scale factor $a(t)$ of the universe , *i.e.* more slowly than t, in a cold matter dominated universe where $a(t) \sim t^{2/3}$.) This implies that if the vacuum energy density is vanishingly small, the part of the universe which we live in now will later probably become the interior of a huge black hole.

However, this does not necessarily mean that mankind is doomed to die inside a black hole[55]. Indeed, as it was explained above, the domains with $\delta\rho/\rho \geq 1$ are formed from inflationary domains which had a size $0(H^{-1})$ at the time where the field φ was of the order φ^*. During inflation each such domain becomes divided into many inflationary domains, some of which have extremely large energy density and should remain inflationary for an unlimitedely long time.

The only possible strategy of survival which we see at the moment is to travel from old domains to new ones, which are displaced near the self-reproducing inflationary domains and always have large enough density of baryons[55]. In the worst case, if we are unable to travel to such distant places ourselves, we can try to send some information about us, our life and our knowledge, and maybe even stimulate development there of such kinds of life, which would be able to receive and use this information. In such a case one would have a comforting thought that even though life in our part of the universe will disappear, we will have some inheritors, and in this sense our existence is not entirely meaningless. (At least it would not be worse than what we have here now.)

In order to check whether such a possibility actually exists, it would be necessary to perform a detailed investigation of the global structure of the universe, and especially of its causal structure, which becomes very nontrivial in the context of inflationary cosmology[54]. One should answer the question whether it is actually possible to travel or to send a signal from our part of the universe to the vicinity of a nearby self-reproducing inflationary domain, and to do it in such a way that after the long time which is required until the signal arrives the density of matter near this domain will be sufficiently large for the existence of life there. At present we do not know of any no-go theorems which would imply that it is generally impossible to work out any successful strategy for survival of life in our part of the universe, if it has vanishing vacuum energy density. However, as we are going to argue now, it does not seem possible that mankind can have an infinitely long lifetime if the vacuum energy density is bigger than some critical value ρ_c, which is extremely small in all realistic versions of the inflationary universe scenario.

Namely, the self-reproducing domains appear in the inflationary universe independently of the existence or nonexistence of a small energy density ρ_v of the vacuum state in which we live now (if it is not negative and as big as $10^{-6}M_p^4$) [54]. However, the causal structure of the universe, the possibility of communication between its different domains and the possibility of the existence of life there depend crucially on the value of ρ_v. For example, if ρ_v is negative and its absolute value is much bigger than the present energy density of matter $\rho_o \sim 10^{-29} g \cdot cm^{-3}$, each domain of the universe after inflation and a sufficiently long subsequent period of expansion behaves as an anti de Sitter space and collapses in a time $t \sim M_P \rho_v^{-1/2} \ll M_P \rho_o^{-1/2} \sim 10^{10}$ years, *i.e.* much earlier than the time necessary for the development of life of our type. Note, that this process occurs locally, at different times in different places of a self-reproducing universe. Therefore one may escape the local collapse by travelling towards the self-reproducing domains. However, since the typical distance between our place and a nearby self-reproducing domain is extremely large, such a journey is possible only if the lifetime of the anti de Sitter domain $t \sim M_P \rho_v^{-1/2}$ is also extremely large and, consequently, the absolute value of ρ_v is extremely small (see below).

If the vacuum energy density is positive and bigger than $10^{-27} g \cdot cm^{-3}$, formation of galaxies of our type becomes impossible[55]. Smaller values of ρ_v do not affect galaxy formation, but do affect the causal structure of the universe. For example, the universe with $\rho_v > \rho_0$ at time $t > 10^{10}$ years behaves as de Sitter space with event horizon of size $H^{-1} \sim 10^{28} cm$ (*i.e.* smaller than the particle horizon $\sim t$ now). In such a universe the distance between the galaxies grows as e^{Ht}, and after a time of the order $H^{-1} < 10^{10}$ years there would remain only one (our own) galaxy inside our event horizon. (The galaxy itself is not stretched by the expansion of the universe since it is a gravitationally bound system with energy density much bigger than ρ_0.) It is well known that in the exponentially expanding universe one can travel or send any information only to those regions which were inside the event horizon at the moment when the information was sent. This means that soon after the beginning of the exponential expansion of the universe with $\rho_v > \rho_0$ there will be no possibility to travel or send information to other galaxies, and it is certainly impossible to send a signal to any self-reproducing inflationary domain, since a typical distance from our place to any such domain is much bigger than H^{-1}. In such a case our galaxy will transform from our cosmic home to a cosmic prison with a finite amount of food and energy supply, which cannot last indefinitely long.

This result can be considerably strengthened in all realistic models of inflation. To do this one should first estimate the typical distance l^* from our place to a nearby inflationary self-reproducing domain. This can easily be done, since there should be many domains in each region of initial size $H^{-1}(\varphi)$ with $\varphi > \varphi^*$, and there should be no domains inside the regions of initial size H^{-1} with $\varphi < \varphi^*$. For the theory with potential $m^2\varphi^2/2$ this gives, according to (7.2), (7.3),

$$l^* \sim H^{-1}(\varphi^*) \exp\left(\frac{2\pi\varphi^{*2}}{M_p^2}\right)\left(\frac{T_R}{T_0}\right) . \tag{7.4}$$

Here $H^{-1}(\varphi^*) = M_P \left(\frac{8\pi V(\varphi^*)}{3}\right)^{-1/2}$, T_R is the reheating temperature of the universe after inflation, T_0 is the present temperature of the blackbody radiation in the universe. The last factor appears due to the additional stretching of the universe when the temperature drops from T_R to T_0. Typically this factor is of the order 10^{30} and $H^{-1}(\varphi^*)$ is of the order 10^{-30} cm. Therefore from (7.3), (7.4) it follows that

$$l^* \sim \exp(2\pi M_P/m)\ cm \sim \exp((2\pi)10^6)\ cm\ . \tag{7.5}$$

In the numerical estimate in (7.5) we use the well-known result that density perturbations generated during inflation in this scenario have an amplitude necessary for the subsequent process of galaxy formation if $m \sim 10^{-6} M_P$. The universe starts to expand exponentially and acquires an event horizon $H^{-1}(\rho_v) = M_P \left(\frac{8\pi\rho_v}{3}\right)^{-1/2}$ during the late stages of its evolution, when the energy density of matter, which decreases as t^{-2}, becomes smaller than the vacuum energy density ρ_v. This occurs at a time which is bigger than the present age of the universe $t_0 \sim 10^{10}$ years by a factor $\left(\frac{\rho_0}{\rho_v}\right)^{1/2}$. At that time the distance l^* from us to the nearby inflationary self-reproducing domain grows by a factor $\left(\frac{\rho_0}{\rho_v}\right)^{1/3}$, since the scale factor of a cold matter dominated universe grows as $t^{2/3}$. The causal connection between us and this region will be possible only if this distance remains smaller than $H^{-1}(\rho_v)$,

$$M_P \left(\frac{8\pi\rho_v}{3}\right)^{-1/2} > M_P \left(\frac{8\pi V(\varphi^*)}{3}\right)^{-1/2} \exp\left(\frac{2\pi\varphi^{*2}}{M_p^2}\right) \left(\frac{T_R}{T_0}\right) \left(\frac{\rho_0}{\rho_v}\right)^{1/3}, \tag{7.6}$$

which implies that causal connection is possible only if ρ_v is

smaller than some critical value ρ_c, where
(7.7)

$$\rho_c \simeq \frac{V^3(\varphi^*)}{\rho_0^2} \left(\frac{T_0}{T_R}\right)^6 \exp\left(-\frac{12\pi M_P}{m}\right) \sim 10^{-10^7} g \cdot cm^{-3} .$$

Here $V(\varphi^*) \sim mM_p^3 \sim 10^{-6}M_p^4$, $\rho_0 \sim 10^{-29} g{\cdot}cm^{-3} \sim 10^{-125}M_p^4$, $T_R \sim 10^{10}\ GeV$. One can easily see that the constraint (7.7) on the value of the vacuum energy density ρ_v is very stringent and practically does not depend on the particular values of $V(\varphi^*), \rho_o$ and T_R.

Returning to the case of a negative vacuum energy, it can be shown by methods similar to those used above that in this case as well causal contact between our part of the universe and a nearby self-reproducing inflationary domain is impossible if $|\rho_v| > \rho_c$, since in this case the time which is necessary for the contact is bigger than the time before the local collapse of our part of the universe.

These results are rather unexpected. It was clear *a priori* that the existence of a nonvanishing vacuum energy may slightly change the conditions of our life , but it was difficult to imagine that the development of life in our part of the universe may depend crucially on the existence or nonexistence of a tiny vacuum energy density $|\rho_v| \geq 10^{-10^7} g \cdot cm^{-3}$.

Of course, the events we are discussing here will occur in a very distant future (at $t \sim 10^{5.10^6}$ years), but it is very difficult not to be curious about our common fate as it can be understood at the present level of the development of science. Another possible application of these results will be discussed in the next section.

8. Cosmological Constant Problem, Universe Multiplication, Baby Universes and Anthropic Principle

The conceptual problems of quantum cosmology are very difficult. In order to solve them one should have a clear understanding of what the universe is, what the physical theory is and even what our life is. Of course, one can simply say that the universe is everything which exists, that the theory is described by the Lagrangian (or Hamiltonian) with some coupling constants, that life appears at late stages of the evolution of the universe and it cannot have any impact on the laws of Nature and on the structure of the universe. However, in the context of quantum cosmology the validity of these common sense statements becomes less clear. One may ask whether the Lagrangian existed before the universe appeared "from nothing," whether the universe existed and evolved in time before it was observed and other nasty questions which one can safely ignore in our everyday life. As will be argued below, an investigation of such questions may lead to important and rather unexpected consequences for the future development of science.

One of the most difficult current problems of elementary particle theory is the cosmological constant problem. This problem was first recognized about 15 years ago, when it was understood that the effective potential $V(\varphi)$ of a classical scalar field φ, being multiplied by $8\pi/M_p^2$, plays the role of the cosmological constant Λ in the Einstein equations[59]. From the comparison of the cosmological observational data with the solutions of the Einstein equations with $\Lambda \neq 0$, it follows that the observed value of the curvature scalar R at present corresponds to an extremely small value of the vacuum energy density, $V(\varphi) \leq 10^{-29} g \cdot cm^{-3}$. The main problem here is to understand the reason why, after a large chain of phase transitions with breaking of various symmetries in unified theories of elementary particles, the vacuum energy in our asymmetric vacuum state exactly (or almost exactly) vanishes. Indeed, a most natural value of the vacuum energy density which could be expected in Kaluza-Klein theories is of the order of the Planck energy density, $\rho_P \sim M_p^4 \sim 10^{94} g \cdot cm^{-3}$, which is at least 125 orders of magnitude greater than the present vacuum energy density $V(\varphi) \leq 10^{-29} g \cdot cm^{-3}$. In grand unified theories a typical value of the vacuum energy density is of the order $10^{80} - 10^{85} g \cdot cm^{-3}$, in the electroweak theory it is $0(10^{25} g \cdot cm^{-3})$. It is very difficult to understand how all these contributions to the total value of the vacuum energy, being summed up, give us something as small as $V(\varphi) \leq 10^{-29} g \cdot cm^{-3}$.

In the last year there were several interesting suggestions for how to solve this problem. Here we would like to discuss some different proposals related to the suggestion[65] that it is possible to solve the cosmological constant problem by considering many universes interacting with each other only globally.

In order to explain what this means, we first consider the original model proposed in ref. 60. This model describes two universes, X and Y, with coordinates x_μ and y_α, respectively $(\mu, \alpha = 0, 1, \ldots, 3)$, and with metrics $g_{\mu\nu}(x)$ and $\bar{g}_{\alpha\beta}(y)$, containing fields $\varphi(x)$ and $\bar{\varphi}(y)$ with an action of the following unusual type:

(8.1)

$$S = N \int d^4x d^4y \sqrt{g(x)} \sqrt{\bar{g}(y)} \times \left[\frac{M_P^2}{16\pi} R(x) + L(\varphi(x)) - \frac{M_P^2}{16\pi} R(y) - L(\bar{\varphi}(y)) \right]$$

Here N is some normalization constant. This action is invariant under general coordinate transformations in each of the universes separately. A novel symmetry of the action is the symmetry under the transformation $\varphi(x) \to \bar{\varphi}(x), g_{\mu\nu}(x) \to$

$\bar{g}_{\alpha\beta}(x)$ and under the subsequent change of the overall sign, $S \to -S$. We call this the antipodal symmetry, since it relates to each other the states with positive and negative energies.

An immediate consequence of this symmetry is invariance under a change of the values of the effective potentials $V(\varphi) \to V(\varphi) + c$, $V(\bar{\varphi}) = V(\bar{\varphi}) + c$, where c is some constant. Consequently, nothing in this theory depends on the value of the effective potentials $V(\varphi)$ and $V(\bar{\varphi})$ in their absolute minima φ_0 and $\bar{\varphi}_0$. (Note, that $\varphi_0 = \bar{\varphi}_0$ and $V(\varphi_0) = V(\bar{\varphi}_0)$ due to the antipodal symmetry.) This is the basic reason why it proves possible to solve the cosmological constant problem in our model.

However, our main reason to invoke this new symmetry was not just to solve the cosmological constant problem. Just as the theory of mirror particles originally was proposed in order to make the theory CP-symmetric while maintaining CP-asymmetry in its observable sector, the theory (8.1) is proposed in order to make the theory symmetric with respect to the choice of the sign of energy. This removes the old prejudice that, even though the overall change of sign of the Lagrangian (*i.e.* both of its kinetic and potential terms) does not change the solutions of the theory, one *must say* that the energy of all particles is positive. This prejudice was so strong, that years ago people preferred to quantize *particles* with *negative energy* as *antiparticles* with *positive energy*, which caused the appearance of such meaningless concepts as negative probability. We wish to emphasize that there is no obstacle to performing a consistent quantization of theories which describe particles with negative energy. Real difficulties appear only when there exist interacting species with both signs of energy. (Actually, this is one of the main problems of quantum cosmology, where one should quantize fields with positive energy, as well as the scale factor a with negative energy, see Sec. 1.) In our case no such problem exists, just as there is no problem of antipodes falling down from the opposite side of the earth. The reason is that the fields $\bar{\varphi}(y)$ do not interact with the fields $\varphi(x)$, and the equations of motion for the fields $\bar{\varphi}(y)$ are the same as for the fields $\varphi(x)$ (the overall minus sign in front of $L(\bar{\varphi}(y))$ does not change the Lagrange equations). Similarly, gravitons from different universes do not interact with each other. However, some interaction between the two universes does exist. Indeed, in our case the Einstein equations are:

(8.2)
$$R_{\mu\nu}(x)-\frac{1}{2}g_{\mu\nu}R(x) = -8\pi GT_{\mu\nu}(x)-g_{\mu\nu}\left\langle\frac{1}{2}R(y)+8\pi GL\big(\bar{\varphi}(y)\big)\right\rangle ,$$

(8.3)
$$R_{\alpha\beta}(y)-\frac{1}{2}\bar{g}_{\alpha\beta}R(y) = -8\pi GT_{\alpha\beta}(y)-\bar{g}_{\alpha\beta}\left\langle\frac{1}{2}R(x)+8\pi GL\big(\varphi(x)\big)\right\rangle$$

Here $T_{\mu\nu}(x)$ is the energy-momentum tensor of the fields $\varphi(x)$, $T_{\alpha\beta}(y)$ is the energy-momentum tensor of the fields $\bar{\varphi}(y)$; the averaging sign means

$$\langle R(x)\rangle = \frac{\int d^4x\sqrt{g(x)}R(x)}{\int d^4x\sqrt{g(x)}}, \tag{8.4}$$

$$\langle R(y)\rangle = \frac{\int d^4y\sqrt{\bar{g}(y)}R(y)}{\int d^4y\sqrt{\bar{g}(y)}}, \tag{8.5}$$

and similarly for $\langle L(x)\rangle$ and $\langle L(y)\rangle$. Thus, the novel feature of the theory (8.1) is the existence of a *global* interaction between the universes X and Y: The integral *over the whole history* of the Y-universe changes the vacuum energy density of the X-universe.

In general, the computation of averages of the type (8.4), (8.5) may be a rather sophisticated problem. Fortunately, however, in the inflationary universe scenario (at least, if the universe is not self-reproducing, see below), this task is rather trivial. Namely, the universe after inflation becomes almost flat and its lifetime becomes exponentially large. In such a case, the dominating contribution to the average values $\langle R \rangle$ and $\langle L \rangle$ comes from the late stages of the evolution of the universe during which the fields $\varphi(x)$ and $\bar{\varphi}(y)$ relax near the absolute minima of their effective potentials. As a result, the average value of $-L(\varphi(x))$ almost exactly coincides with the value of the effective potential $V(\varphi)$ in its absolute minimum at $\varphi = \varphi_0$, and the average value of the curvature scalar $R(x)$ coincides with its value during the late stages of the evolution of the universe, when the universe transforms to the state corresponding to the absolute minimum of $V(\varphi)$. Similar results are valid for the average values of $-L(\bar{\varphi}(y))$ and of $R(y)$ as well. In such a case one can easily show[60] that during the late stages of the evolution of the universe, when the fields $\varphi(x)$ and $\bar{\varphi}(y)$ relax near the absolute minima of their effective potentials, the *effective* cosmological constant automatically vanishes,

$$R(x) = -R(y) = \frac{32}{3}\pi G[V(\varphi_0) - V(\bar{\varphi}_0)] = 0 \qquad (8.6)$$

If the universe is self-reproducing, then one may encounter some difficulties when computing the averages (8.4), (8.5), since they may become infrared-divergent and the result of the computation may depend on the cut-off. This question is not yet completely investigated due to the very complicated large-scale structure of the self-reproducing universe. However, one can easily avoid such questions in theories in which $V(\varphi)$ grows rapidly enough at $\varphi \geq \varphi^*$, since there will be no universe self-reproduction in such theories. Another problem is that the integral over d^4y in (8.1) renormalizes the effective Planck constant, and one should take a very small normalization factor $[N \sim \exp(-\lambda^{1/3})$ in the $\lambda\varphi^4$ theory$]$ in order to compensate this renormalization. Another possibility is that in constructing the quantum theory in a doubled universe, one should simply do it in each of the noninteracting universes separately, without taking into account the abovementioned renormalization of N. Note also that the mechanism for the cancellation of the cosmological term suggested above works independently of the value of N.

The model considered above can be easily extended. For example, one can consider not only a doubled universe, but also a multiuniverse, with the Lagrangian being a sum of different Lagrangians of different theories of different fields living in different disconnected universes.

An extremely interesting generalization of this construction was proposed by Coleman[61], Giddings and Strominger[62] and Banks[63], which is based on previous works by Hawking[64], Lavrelashvili, Rubakov and Tinyakov[65] and Giddings and Strominger[66] on the wormholes and coherence loss in quantum gravity, and also on the proposal of Baum and Hawking concerning the cancellation of the cosmological term[67,68]. The main idea is that our universe can be split into disconnected pieces by quantum gravity effects. Baby universes created from the parent universe can carry from it an electron-positron pair, or some other combinations of particles and fields, unless this is forbidden by conservation laws. Such a process can occur in any place in our universe. Many ways were suggested to describe such a situation. The simplest one is to say that the existence of baby universes leads to a modification of the effective Hamiltonian density[61–63]

$$\mathcal{H}(x) = \mathcal{H}_0(\varphi(x)) + \sum \mathcal{H}_i[\varphi(x)]A_i \tag{8.7}$$

The Hamiltonian (8.7) describes the fields $\varphi(x)$ on the parent universe at distances much greater than the Planck scale. $\mathcal{H}_0$ is the part of the Hamiltonian which does not involve topological fluctuations. $\mathcal{H}_i(\varphi)$ are some local functions of the fields φ, and A_i are combinations of creation and annihilation operators for the baby universes. These operators do not depend on x since the baby universes cannot carry away momentum. Coleman argues [61] that the demand of locality, on the parent universe,

$$[\mathcal{H}(x), \mathcal{H}(y)] = 0 \tag{8.8}$$

for spacelike separated x and y, implies that the operators A_i must all commute. Therefore, they can be simultaneously diagonalized by the "$\alpha - state$":

$$A_i|\alpha_i >= \alpha_i|\alpha_i > \ . \tag{8.9}$$

If the state of the baby universe is an eigenstate of the A_i, then the net effect of the baby universes is to introduce infinite number of undetermined parameters (the α_i) into the effective Hamiltonian (8.7): one can just replace the operators A_i by their eigenvalues. If the universe initially is not in an A_i eigenstate, then, nevertheless, after a series of measurements the wave function soon collapses to one of the A_i eigenstates[61–63].

This gives rise to an extremely interesting possibility related to the basic principles of physics. We were accustomed to believe that the main purpose of physics is to discover the Lagrangian (or Hamiltonian) of the theory which correctly describes our world. However, the question arises: if our universe did not exist sometime in the distant past, in which sense could one speak about the existence of the laws of Nature which govern the universe? We know, for example, that the laws of our biological evolution are written in our genetic code. But where were the laws of physics written at the time when there was no universe (if there was such time)? The possible answer now is that the final structure of the (effective) Hamiltonian becomes fixed only after measurements are performed, which determine the values of coupling constants in the state in which we live with better and better precision. Different effective Hamiltonians describe different laws of physics in different (quantum) states of the universe, and by making measurements we reduce the variety of all possible laws of physics to those laws which are valid in the (classical) universe where we live.

Another important extension of the ideas of refs. 61-63 is connected with the possibility that the wave function of the universe depends on the values of the coupling constants in each quantum state, and that it can be peaked near some particular values of these constants. The most interesting application is the possible explanation of the vanishing of the cosmological constant suggested by Coleman[69]. The main idea is closely related to the previous suggestion by Hawking[68] based upon the Hartle-Hawking wave function of the universe (1.1), (1.4). According to Hawking[68], if the cosmological constant can take on any given value in our universe, then the probability to find ourselves in the universe with the cosmological constant $\Lambda = 8\pi V(\varphi)/M_p^2$ would be given by (1.4)

$$P(\Lambda) \sim \exp\left(-2S_E(\Lambda)\right) = \exp\left(\frac{3\pi M_p^2}{\Lambda}\right). \tag{8.10}$$

In the baby universe theory discussed above the cosmological constant, like other constants, actually can take different values. However, in this theory one should not only take into account one-universe Euclidean configurations. Rather one should sum over all configurations of babies and parents (connected by Euclidean wormholes), which finally gives[69]

$$P(\Lambda) \sim \exp\left(\exp\left(\frac{3\pi M_p^2}{\Lambda}\right)\right). \tag{8.11}$$

From (8.11) it follows that it is most probable to live in a universe with $\Lambda = 0$.

Is this conclusion reliable? It is very difficult to give an answer to this question. As we have discussed in Sec. 3, it was possible to justify the use of the wave function (1.1), (1.4) only in those cases where some kind of a stationary distribution of $V(\varphi)$ may actually exist, and we have found no such distribution in realistic cosmological models. It might be that the distribution of the values of coupling constants α_i is stationary (it should be) and is described by equations of the type of (1.5) or (8.10), (8.11), but we cannot justify this statement by the use of the stochastic approach, as was done in Sec. 3. In particular, the stochastic approach to the baby universe formation[54] gives the results (6.15), (6.16), which are different from those obtained by euclidean methods[64–66]. On the other hand, if one uses the tunneling wave function (1.6) instead of (1.4), it may seem that the typical value of the vacuum energy will be $0(M_P^4)$, which is undesirable. However, there may exist a way to overcome this difficulty.

Namely, in ref. 69 only the simplest closed Friedmann universes with scale factor $a(t)$ were considered. For such universes there exists a barrier between $a = 0$ and $a = H^{-1}(\Lambda)$ due to the term $\sim -a^2$ in eq. (1.3). This term disappears in a flat Friedmann universe, and if such a universe is compact, the corresponding Euclidean action becomes zero. An example of such a universe is a 3-torus with metric

$$ds^2 = dt^2 - a_i^2(t)dx_i^2 \tag{8.12}$$

where the points x_i and $x_i + L$ are identified[6]. After inflation all local properties of such a universe will be indistinguishable from the properties of the ordinary Friedmann universe. Since there is no exponential suppression of the contribution of such universes to the wave function of the universe (due to the vanishing of the corresponding Euclidean action $S_E(\Lambda)$), such universes may play an important role in the calculation of $P(\Lambda)$. It might be possible that with taking into account such configurations there will be no exponential suppression of the probability to live in the universe with any particular value of Λ, independently of the choice of the wave function (1.4) or (1.6). In such a case one may consider other mechanisms which ensure the smallness of the cosmological constant, *e.g.* the mechanism suggested by Banks[63].

At the moment we do not know which of these mechanisms, if any, makes the vacuum energy vanishingly small. It would be very interesting to understand how these mechanisms work by considering some nontrivial models. Let us consider a theory which has many different absolutely stable vacuum states $\varphi_1, \varphi_2, \ldots$ with different energy densities $V(\varphi_1), V(\varphi_2), \ldots$. A simple example of such a theory is the supersymmetric $SU(5)$ model. Its effective potential $V(\Phi, H)$ has many different minima[70] with different energy density (after supergravity effects have been taken into account), which nevertheless correspond to absolutely stable vacuum states[71].

In accordance with our previous investigation, in a self-reproducing inflationary universe exponentially large domains containing all possible vacua $\varphi_1, \varphi_2, \ldots$ are produced. If the difference between $V(\varphi_i)$ is much smaller than M_P^4, then the *number* of domains with different fields φ_i should be of the same order of magnitude, whereas the main part of the *volume* of the universe would be occupied by domains with a maximal possible value of $V(\varphi_i)$. In such a case one may wonder in which of these absolutely stable vacua the cosmological constant should vanish, and why we live just in this vacuum and not in another one? This question is not trivial, since the choice of a particular vacuum state $|\alpha_i>$ shifts the value of Λ by a *constant* uniformly in the whole universe, so it cannot make Λ vanish in all minima of $V(\varphi)$ simultaneously.

A possible answer to this question is the following[72]. If one of the mechanisms of the cancellation of the cosmological constant based on baby universe theory is operative, then the wave function of the universe probably must have not one, but several peaks, at different values of Λ. One of these peaks corresponds to $\Lambda(\varphi_1) = 0, \Lambda(\varphi_i) \neq 0$ for $i \neq 1$, another corresponds to $V(\varphi_2) = 0,\ V(\varphi_i) \neq 0$ for $i \neq 2$, etc. When making measurements, an observer finds himself in a universe corresponding to one of these peaks. This means that he lives in a quantum state with, say, $V(\varphi_2) = 0, V(\varphi_1), V(\varphi_3), ... \neq 0$. Any observer of our type can live only in the universe with $-10^{-29} g \cdot cm^{-3} \leq V(\varphi) \leq 10^{-27} g \cdot cm^{-3}$ [58]. Since the typical difference between $V(\varphi)$ in all realistic theories is many orders of magnitude bigger than $10^{-27} g \cdot cm^{-3}$, an observer of our type can find himself only in the domain of the universe containing the field $\varphi = \varphi_2$, independently of the value of its volume, as compared with the volume occupied by domains corresponding to other minima of $V(\varphi)$.

This answer is based on the use of the Anthropic Principle[73,74]. One should note, that until very recently, the general attitude of physicists to the Anthropic Principle was rather skeptical, to say the least. It was believed that the weak, strong and electromagnetic interactions are the same in all parts of our universe, that the fundamental constants of Nature are universal and it is meaningless to discuss the possibility of life in a universe of a different type. This attitude was somewhat changed first when it was understood that in accordance with the chaotic inflationary universe scenario the universe should contain an exponentially large (or maybe even infinitely large) number of exponentially large domains (mini-universes), in which all kinds of symmetry breaking and all kinds of compactification, which are possible in a theory of a given type, are actually realized. In other words, the universe becomes divided into many exponentially large domains with different types of low-energy physics and maybe even with different dimensionality inside each of them. This made it possible to get a justification of a version of the Weak Anthropic Principle : if many exponentially large domains with the low-energy physics of our type do exist in the universe described by a given theory, then it is quite natural that we live in one of such domains rather than in a domain where life of our type is impossible[38]. With the invention of the theory in which many universes interact with each other only globally[60], and especially with the development of the baby universe theory[61–63] the situation changed even more. As will be argued below, in the context of the baby universe theory it is also possible to justify the Strong Anthropic Principle.

According to the baby universe theory, we may live in different quantum states of the universe $|\alpha_i>$, corresponding to different choices of the "fundamental constants," such as the vacuum energy ρ_v (cosmological constant $\Lambda = 8\pi\rho_v/M_p^2$), gravitational constant $G = M_p^{-2}$, fine structure constant α, etc. By measuring the coupling constants we actually determine which quantum state $|\alpha_i>$ we live in now, and after that we cannot "jump" to another quantum state with other constants. That is why we are used to believe that we live in a classical universe with well- determined and unchanging laws of physics.

It would be very desirable to understand why we live in a universe given by a quantum state $|\alpha_i>$ with an extremely small (or even vanishing) cosmological constant, with $M_P \sim 10^{19}$ GeV, $\alpha = 1/137$, etc. As we have already mentioned, a possible idea is to compute the wave function of the universe and then to prove that it is entirely concentrated near some particular values of the coupling constants[63,69]. However, the validity of the euclidean approach to quantum cosmology used in ref. 69 and in most of other papers on this problem is questionable. It is also quite possible that the wave function of the universe actually does not have any sharp peak at $\Lambda = 0$ [75,76], and the possibility that it is peaked at some particular values of e, m_e or G is even much less clear. Moreover, it is not quite clear whether it is the wave function of the whole universe that is to be computed. It is known that in quantum mechanics only those statements can make sense that can be formulated in a concrete operational way. For example, at the classical level one can speak of the age of the universe t. However, the essence of the Wheeler-DeWitt equation for the wave function of the universe is that this wave function *does not depend on time*, since the total Hamiltonian of the universe, including the Hamiltonian of the gravitational field, vanishes identically[1,2]. Therefore if one would wish to describe the evolution of the universe with the help of its wave function, one would be in trouble. The resolution of this paradox is rather instructive[2]. The notion of evolution is not applicable to the universe as a whole since there is also no external observer with respect to the universe, and there is no external clock which does not belong to the universe. However, we do not actually ask why the universe *as a whole* evolves in the way we see it. We are just trying to understand our own experimental data. Thus, a more precisely formulated question is why *we see* the universe evolving in time in a given way. In order to answer this question one should first divide the universe into two main pieces: i) an observer with his clock and other measuring devices and ii) the rest of the universe. Then it can be shown that the wave function of the rest of the universe depends on the state of the clock of the observer, *i.e.* on his "time"[2]. This time dependence is in some sense "objective," which means that the results obtained by different (macroscopic) observers living in the same quantum state of the universe and using sufficiently good (macroscopic) measuring apparatus agree with each other[77].

This example teaches us an important lesson. By investigating the wave function of the universe *as a whole* one sometimes gets information which has no direct relevance to the observational data, *e.g.* that the universe does not evolve in time. (Moreover, it is rather difficult to give any probabilistic interpretation to the wave function of the universe as a whole, since typically it is not normalizable.) In order to describe the universe *as we see it* one should divide the universe into several macroscopic pieces and calculate a conditional probability to observe it in a given state under an obvious condition that the observer and his measuring apparatus exist. (A similar approach was used *e.g.* in ref. 4.) This simple condition, however, is very restrictive. In order to understand its meaning one should remember the famous Einstein-Podolsky-Rosen experiment, where two fermions (or photons) are produced by a decay of a spin zero particle. By measuring the spin of one of the particles an observer instantaneously determines the spin of another particle even if this particle and the observer are not in a causal contact at the moment of the measurement. This apparent paradox is resolved if one takes into account that prior to the measurement one should describe both particles by a wave function for the pair of particles rather than by separate wave functions for each of them. The knowledge of this wave function (the fact that it corresponds to a state with zero angular momentum) and the knowledge of the spin of one of these particles makes it possible to determine the spin of another particle without making any causal contact with it, simply because there is a strong correlation between spins of particles in a system with vanishing total angular momentum. For example, the conditional probability to find the second particle in a state with spin $s = -1/2$ is equal to unity under the condition that the first particle has spin $s = +1/2$.

Similarly, any scientific exploration of the properties of the universe starts with the division of the universe into two subsystems (the observer with his measuring devices and the rest of the universe). Due to the strong correlation between the properties of these two subsystems, an investigation of our own properties, the properties of our environment and our measuring devices makes it possible to say a lot about the properties of the rest of the universe such as the vacuum energy, the values of some coupling constants etc.

It is also important that in the EPR experiment one can choose different apparatus, *e.g.* the one which measures a linear polarization of the first photon, or the one which measures its circular polarization. After the measurement, the second photon will be either linearly or circularly polarized, depending on our choice of measuring device. Similarly, many kinds of life and many types of measuring devices may exist, but what we are studying is the correlation of *our* kind of life, environment and measuring devices with the properties of the rest of the universe. But this means that in the context of the baby universe theory one can actually make sense of the Strong Anthropic Principle and try to explain why the electromagnetic coupling constant, the gravitational coupling constant and some other parameters of the theory have those particular values which we know: With other values of the coupling constants the existence of life *of our type* and of the measuring devices we use would be impossible[78,79]. (Note, however, that in this interpretation the Strong Anthropic Principle has nothing in common with designing the universe for the benefit of human beings or with changing the whole universe by our own will, just as the determination of the properties of the second particle in the EPR experiment and the dependence of these properties on our choice of measuring device has nothing to do with the acausal action at a distance.)

If this idea is correct, then there is no need to compute the most probable values of the cosmological constant and of the gravitational constant by investigating the wave function of the universe as a whole, and it is not surprising that some of the results of the corresponding investigation look as counterintuitive as the statement that the universe as a whole does not depend on time. Rather one should investigate the probability to measure some particular eigenvalues of the operators corresponding to the "constants" Λ, e, m_e, G in the quantum state of the universe in which we live now, under the obvious condition that we can live and perform our measurements here. In some cases (*e.g.* in the case of the constants e, m_e, G) the corresponding investigation is rather simple and straightforward. For example, according to ref. 73, an increase of m_e by more than 2.5 times would make impossible the existence of atoms. An increase of e by more than 30% would lead to an instability of protons and nuclei. A change of the strong interaction constant by more than 10 % would lead to the absence of heavy nuclei and of hydrogen in the universe. An increase or decrease of the weak interaction constant by an order of magnitude would lead to the absence of complex elements or of hydrogen respectively. An increase of the gravitational constant G by an order of magnitude would make the lifetime of the Sun so small that no biological molecules would appear on the Earth. An even bigger increase of G would lead to an extremely efficient nucleosynthesis and to the absence of hydrogen in the universe[73,74]. A smaller value of G would slow the expansion of the universe. In such a universe the departure from thermal equilibrium during the process of baryosynthesis would be small, this process would be inefficient and the universe now would be practically empty.

This list can be easily continued, but we hope that the results discussed above clearly demonstrate that the correlation between the properties of observers and the properties of the quantum state of the rest of the universe is very strong indeed. Moreover, most of the results discussed above are not so much *anthropic*, they just show that a small deviation from the present values of coupling constants would make impossible the existence of all measuring devices we use. In the context of the baby universe theory this means that the distribution of the conditional probability to make measurements in a quantum state with given values of weak, strong, gravitational and some other "fundamental constants" proves to be sharply peaked near the actual values of these constants we have measured in the quantum state of the universe in which we live now. This peak in the probability distribution should be especially narrow in unified theories of all fundamental interactions, where the values of different coupling constants are strongly correlated due to the underlying symmetry of the theory. This indicates that if one wishes to get an order-of-magnitude estimate of the coupling constants, in some cases one can do it without computing these constants in the context of the "big fix" paradigm[69], and if one still wishes to compute them (which, of course, is a good idea), then one should compute the amplitudes corresponding to the actual experimental environment rather than the amplitudes corresponding to the metaphysical "universe in itself" investigated by an imaginary external observer. (The discussion of the time-independence of the wave function of the universe shows that in the context of quantum cosmology the subtle difference between the amplitudes of these two types can be crucially important.)

Now let us return to the discussion of the cosmological constant problem. The possibility of explaining the small value of the vacuum energy density $|\rho_v| < 10^{-29} g \cdot cm^{-3}$ with the help of the Anthropic Principle is much less clear, but is not hopeless. From the results of ref. 58 it follows that formation of galaxies and appearance of life of our type is possible only if $-10^{-29} g \cdot cm^{-3} < \rho_v < 10^{-27} g \cdot cm^{-3}$. In the context of the baby universe theory this means that the conditional probability to measure the vacuum energy density ρ_v is concentrated in the interval between $-10^{-29} g \cdot cm^{-3}$ and $10^{-27} g \cdot cm^{-3}$ under the condition that the measurement is made by an observer of our type. This result could be strengthened up to the desirable constraint $|\rho_v| < 10^{-29} g \cdot cm^{-3}$ if it would be possible to prove that life on the Earth has an extragalactic origin. (Any communication between galaxies would be hampered by the exponential expansion of the universe with $\rho_v > 10^{-29} g \cdot cm^{-3}$.) Unfortunately, at the moment we have no evidence in favour of this extravagant hypothesis. (Of course, one can turn this argument upside down and say that the vanishing of the cosmological constant can be considered as an argument in favour of the extragalactic origin of life. However, this will be true only if no other solution of the cosmological constant problem is found.)

Another possibility is related to the results obtained in the previous section. It may not be unreasonable to assume that the conditional probability to be born in the universe in a quantum state with a given value of ρ_v is proportional to the total volume of those parts of the universe with this value of ρ_v where life of our type can appear. From the investigation performed in Section 7 it follows that in the quantum states with $|\rho_v| < 10^{-29} g \cdot cm^{-3}$ the total volume of the parts of the universe where life can exist is much bigger than in the states with $|\rho_v| < 10^{-29} g \cdot cm^{-3}$, and it may be especially big in the states with $|\rho_v| < 10^{-10^7} g \cdot cm^{-3}$. If there is some relation between the total volume of the parts of the universe where life can exist and the volume where life of our type can appear, then one can argue that the probability that we measure the vacuum energy density ρ_v in our quantum state of the universe $|\alpha_i >$ is peaked in the region $|\rho_v| < 10^{-10^7} g \cdot cm^{-3}$. This would be more than enough to solve the cosmological constant problem.

Unfortunately, this line of reasoning, which may lead to a justification of the so-called Final Anthropic Principle[74] in its strongest form, is still far from being well elaborated. There are many obvious and less obvious objections which can be raised against the arguments suggested above. It is quite possible that the cosmological constant problem eventually will be solved in quite a different way, *e.g.* by some modification of ideas proposed in refs. 60, 63, 69. However, the efficiency of the Anthropic Principle for the determination of many other coupling constants in the context of the baby universe theory and the existence of a strong correlation between the value of the cosmological constant and the lifetime of mankind makes it very tempting to continue the investigation of the possibilities discussed above. At the very least, what we have obtained already is a surprising result that the existence of the vacuum energy density as small as $10^{-10^7} g \cdot cm^{-3}$ would be fatal for the future evolution of life in our part of the universe. This is the first indication of the potential practical importance of recent attempts to investigate the cosmological constant problem. Unfortunately (or, may be, fortunately), it may take as much as $10^{5.10^6}$ years until the significance of the current work on this problem will be fully appreciated. Hopefully, within this time we will be able to elaborate either a successful strategy of survival or a more philosophical attitude to our future.

Chapter 3: References

[1] Wheeler, J. A., in '*Battelle Rencontres*,' ed. C. DeWitt and J. A. Wheeler (Benjamin, New York, 1968).

[2] DeWitt, B. S., *Phys. Rev.* 160, 1113 (1967).

[3] Hartle, J. B., and Hawking, S. W., *Phys. Rev.* D28, 2960 (1983).

[4] Hawking, S. W., and Page, D., *Nucl. Phys.* B264, 185 (1986).

[5] Linde, A. D., *Sov. Phys. - JETP* 60, 211 (1984); *Lett. Nuovo Cimento* 39, 401 (1984).

[6] Zeldovich, Ya. B., and Starobinsky, A. A., *Pis'ma Astron. Zh.* 10, 323 (1984).

[7] Rubakov, V. A., *Phys. Lett.* 148B, 280 (1984).

[8] Vilenkin, A., *Phys. Rev.* D30, 509 (1984); *Phys. Rev.* D33, 3560 (1986).

[9] Vilenkin, A., and Ford, L., *Phys. Rev.* D26, 1231 (1982); Linde, A. D., *Phys. Lett.* 116B, 335 (1982); Starobinsky, A. A., *Phys. Lett.* 117B, 175 (1982).

[10] Bogolubov, N. N., and Bogolubov, N. N., Jr., *Introduction to Quantum Statistical Mechanics* (World Scientific, Singapore, 1982).

[11] Vilenkin, A., *Nucl. Phys.* B226, 527 (1983); Brandenberger, R., *Nucl. Phys.* B245, 328 (1984).

[12] Starobinsky, A. A., in *Fundamental Interactions* (MGPI Press, Moscow, 1983), p. 55; Starobinsky, A. A., in *Current Topics in Field Theory, Quantum Gravity and Strings*, Lecture Notes in Physics, eds. H. J. de Vega and N. Sanchez (Springer-Verlag, Heidelberg, 1986), vol. 246, p. 107.

[13] Goncharov, A. S., and Linde, A. D., *Fiz. Elem. Chastits At. Yadra* 17, 837 (1986) [*Sov. J. Part. Nucl.* 17, 369 (1986)].

[14] Mukhanov, V. F., and Chibisov, G. V., *JETP Lett.* 33, 532 (1981); *Sov. Phys. - JETP* 56, 258 (1982); Hawking, S. W., *Phys. Lett.* 115B, 295 (1982); Starobinsky, A., *Phys. Lett.* 117B, 175 (1982); Guth, A. H., and Pi, S.-Y., *Phys. Rev. Lett.* 49, 1110 (1983); Bardeen, J., Steinhardt, P., and Turner, M., *Phys. Rev.* D28, 679 (1983).

[15] Gibbons, G. W., and Hawking, S. W., *Phys. Rev.* D15, 2738 (1977).

[16] Hawking, S. W., and Moss, I. G., *Phys. Lett.* 110B, 35 (1982); *Nucl. Phys.* B224, 180 (1983).

[17] Linde, A. D., *Mod. Phys. Lett.* 1A, 81 (1986).

[18] Linde, A. D., *Phys. Lett.* 175B, 395 (1986).

[19] Linde, A. D., in *Proceedings of the Nobel Symposium on Elementary Particle Physics*, Physica Scripta 34 (1986),

to be published.

[20] Linde, A. D., in *300 Years of Gravitation*, Hawking, S. W., and Israel, W., eds., (Cambridge University Press, Cambridge, 1987).

[21] Goncharov, A. S., and Linde, A. D., *Zh. Eksp. Teor. Fiz.* 92, 1137 (1987).

[22] Guth, A. H., *Phys. Rev.* D23, 347 (1981).

[23] Gott, J., *Nature* 295, 304 (1982); Sato, K., Kodama, H., Sasaki, M., and Maeda, K., *Phys. Lett.* 108B, 35 (1982); Guth, A.H., and Weinberg, E., *Nucl. Phys.* B212, 321 (1983).

[24] Steinhardt, P. J., in *The Very Early Universe*, Hawking, S. W., Gibbons, G. W., and Siklos, S., eds., (Cambridge University Press, Cambridge, 1983) p. 251.

[25] Linde, A. D., "*Non-singular Regenerating Inflationary Universe*," Cambridge University preprint (1982).

[26] Vilenkin, A., *Phys. Rev.* D27, 2848 (1983).

[27] Ma, S.-K., *Statistical Mechanics* (World Scientific, Singapore, 1985).

[28] Graziani, F. R., and Olynyk, K. O., Fermilab preprint Pub-85/175- T (1985).

[29] Bardeen, J. M., and Bublik, G. J., *Classical and Quantum Gravity* 4, 573 (1987).

[30] Pollock, M. D., Trieste preprint IC/86/65 (1986).

[31] Linde, *Phys. Lett.* 162B, 281 (1985); *Prog. Theor. Phys. Suppl.* 85, 279 (1985).

[32] Zeldovich, Ya. B., *Pis'ma Astron. Zh.* 7, 579 (1981).

[33] Starobinsky, A. A., *Phys. Lett.* 91B, 99 (1980).

[34] Mukhanov, V. F., *JETP Lett.* 41, 493 (1985).

[35] Chibisov, G. V., and Mukhanov, V. F., Lebedev Institute preprint N154 (1983).

[36] Bardeen, J., *Phys. Rev.* D22, 1882 (1980).

[37] Linde, A. D., *Rep. Prog. Phys.* 47, 925 (1984).

[38] Linde, A. D., *Particle Physics and Inflationary Cosmology* (Gordon and Breach, New York, to be published.)

[39] Rubakov, V. A., Sazhin, M. V., and Veryaskin, A. V., *Phys. Lett.* 115B, 189 (1982); Hawking, S. W., *Phys. Lett.* 150B, 339 (1985).

[40] Turner, M. S., in *Proc. of the Cargese School on Fundamental Physics and Cosmology*, Audouse, J., and Tran Thanh Van, J., eds., (Editions Frontiers: Gif-sur-Yvette, 1985).

[41] Enquist, K., and Maalampi, J., *Phys. Lett.* 180B, 14 (1986).

[42] Smolin, L., *Phys. Lett.* 93B, 95 (1980).

[43] Spokoiny, B., *Phys. Lett.*, 147B, 39 (1984); Accetta, F. S., Zoller, D. J., and Turner, M. S., *Phys. Rev.* D31, 3046 (1985).

[44] Futamase, T., and Maeda, K., preprint UTAP-62/87 (1987); Frieman, J., to be published.

[45] Linde, A. D., *Phys. Lett.* 93B, 394 (1980).

[46] Starobinsky, A. A., *Sov. Astronomy Lett.* 7, 36 (1981); Pollock, M., *Phys. Lett.* 108B, 386 (1982).

[47] Kofman, L. A., Linde, A. D., and Starobinsky, A. A., *Phys. Lett.* 157B, 361 (1985).

[48] Linde, A. D., *Phys. Lett.* 129B, 177 (1983).

[49] Linde, A. D., *Phys. Lett.* 131B, 330 (1983).

[50] Linde, A. D., and Zelnikov, M., to be published.

[51] Randjbar-Daemi, S., Salam, A., and Strathdee, J., *Nucl. Phys.* B214, 491 (1983).

[52] Mandelbrot, B. B., *Fractals: Form, Chance and Dimensions* (Freeman, San Francisco, 1977); Aryal, M. and Vilenkin, A., *Phys. Lett.* B199, 351 (1987).

[53] Farhi, E. and Guth, A. H., *Phys. Lett.* B183, 149 (1987).

[54] Goncharov, A. S., Linde, A. D. and Mukhanov, V. F., *Intern. J. Mod. Phys.* A2, 561 (1987).

[55] Linde, A. D., *Phys. Lett.* B211, 29 (1988).

[56] Zeldovich, Ya. B. and Novikov, I. D., *Sov. Astron.* 10, 602 (1967); Hawking, S. W., *Mon. Not. R. Astron. Soc.* 152, 75 (1971).

[57] Polnarev, A. G., Khlopov, M. Yu., *Sov. Phys.-Uspekhi* 28, 213 (1985).

[58] Weinberg, S., *Phys. Rev. Lett.* 59, 2607 (1987); *Rev. Mod. Phys.* 61, 1 (1989).

[59] Linde, A. D., *JETP. Lett.* 19, 183 (1974); Veltman, M., Rockefeller University preprint (1974); *Phys. Rev. Lett.* 34, 77 (1975); Dreilein, J., *Phys. Rev. Lett.* 33, 1243 (1974).

[60] Linde, A. D., *Phys. Lett.* B200, 272 (1988).

[61] Coleman, S., *Nucl. Phys.* B307, 867 (1988).

[62] Giddings, S. and Strominger, A., *Nucl. Phys.* B307, 854 (1988).

[63] Banks, T., *Nucl. Phys.* B309, 493 (1988).

[64] Hawking, S. W., *Phys. Lett.* B195, 277 (1987); *Phys. Rev.* D37, 904 (1988).

[65] Lavrelashvili, G. V., Rubakov, V. A. and Tinyakov, P. G., *JETP Lett.* 46, 167 (1987); *Nucl. Phys.* B299, 757 (1988).

[66] Giddings, S. and Strominger, A., *Nucl. Phys.* B306, 890 (1988).

[67] Baum, E., *Phys. Lett.* B133, 185 (1983).

[68] Hawking, S. W., *Phys. Lett.* B134, 403 (1984).

[69] Coleman, S., *Nucl. Phys.* B310, 643 (1989).

[70] Dragon, N.V., *Phys. Lett.* B113, 288 (1982); Frampton, P. H. and Kephart, T. W., *Phys. Rev. Lett.* 48, 1237 (1982); Buccella, F., Derendinger, J. P., Ferrara, S. and Savoy, S. A., *Phys. Lett.* B115, 375 (1982).

[71] Weinberg, S., *Phys. Rev. Lett.* 48, 1776 (1982).

[72] Rubakov, V.A., private communication.

[73] Rosental, I. L., *Big Bang, Big Bounce: How Particles and Fields Drive Cosmic Evolution*, (Springer, Berlin, 1988).

[74] Barrow, J., and Tipler, F., *The Anthropic Cosmological Principle*, (Oxford University Press, Oxford, 1986).

[75] Polchinski, J., *Phys. Lett.*, to be published.

[76] Fischler, W., Klebanov, I., Polchinski, J. and Susskind, L., Stanford University preprint, (1989).

[77] Everett, H., *Rev. Mod. Phys.* 29, 454 (1957); DeWitt, B. S. and Graham, N., *The Many-Worlds Interpretation of Quantum Mechanics*, (Princeton University Press, Princeton, 1973).

[78] Rubakov, V.A. and Shaposhnikov M.E., *Mod. Phys. Lett.* 4A, 107 (1989).

[79] Linde, A.D., CERN preprint TH.5389/89 (1989); to be published in *Phys. Lett.*.

APPENDIX:

MORE ON PARTICLE PHYSICS AND THE CREATION OF THE UNIVERSE

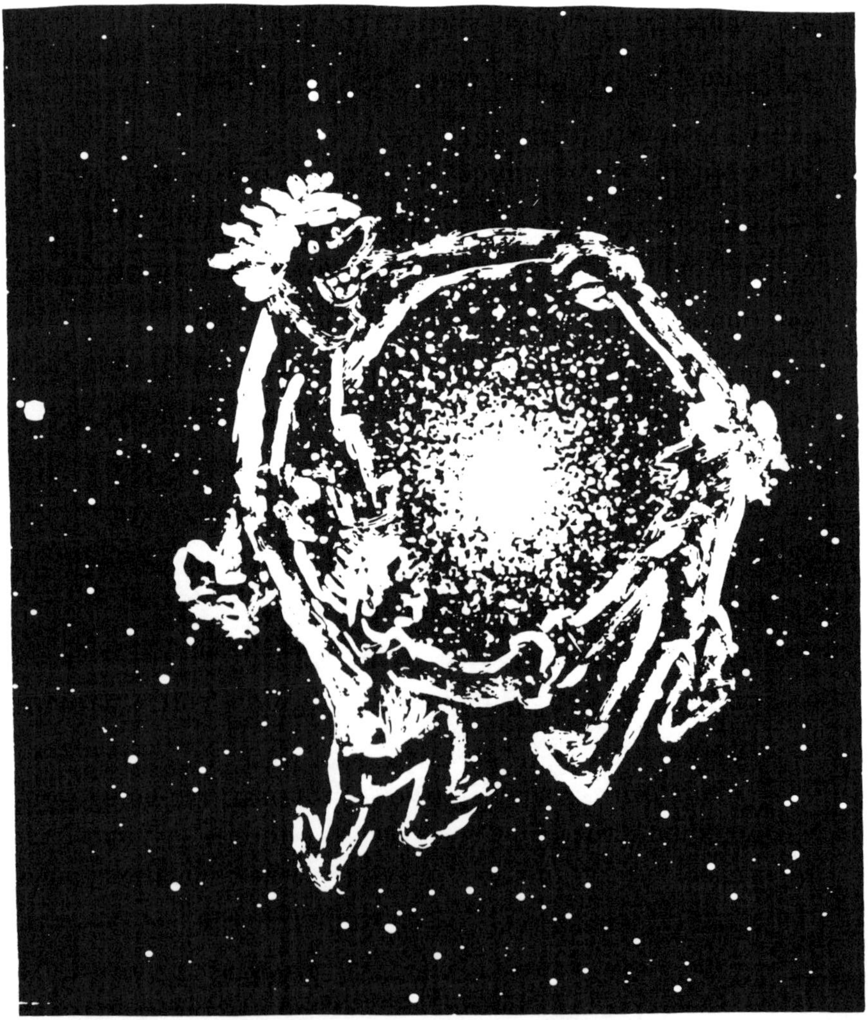

Fig. 1. One of the main problems in giving lectures is to make complicated material look simple, such that, as we say in Russia, even academics can understand it. That is why I have drawn some cartoons with which I illustrate various features of inflationary cosmology in my talks at different conferences, starting from the first conference on inflation held in Cambridge, England, 1982. The collection of illustrations gradually grew, and it was recommended to me that I add these illustrations to the book in order to show that we now not only have a scenario, but also decorations. I recommend that the readers look at the illustrations *after* reading Chapter 1. One cannot study physics by looking at cartoons, but, on the other hand, one should have some fun from studying cosmology.

A short happy life...

Fig. 2. An intensive interplay between cosmology and particle physics started in the middle of the 70s. At first everyone was very happy. Those who studied particle physics gave cosmologists many nice presents: massive neutrinos, which may be helpful for galaxy formation, unstable protons, which are necesary for the explanation of the absence of antimatter in our part of the universe, etc.

„Elementary particle physics is incompatible with cosmology"

(Anonymous, 1979)

Fig. 3. Unfortunately, this happy life was very short. In 1978-1979 it was realized that according to grand unified theories, there now should be too many primordial magnetic monopoles produced during the very early stages of the evolution of the hot universe. The discrepancy between theoretical predictions and the cosmological observational data was more than 15 orders of magnitude. After many unsuccessful attempts to solve the primordial monopole problem, some people began to say that elementary particle physics is incompatible with cosmology. Experts in elementary particle physics repeated the remark by Landau that astrophysicists are always in error and never in doubt. Cosmologists, in turn, argued that there is no reason to worry about grand unification, which may occur or may not occur on an energy scale 12 orders of magnitude bigger than the energy scale which can be tested with the help of the existing accelerators. It seemed possible that one could easily overcome all difficulties by choosing some better theory of elementary particles. However, the new theories brought new problems and did not resolve the previous ones.

Green
Schwarz
O(32)

Fig. 4. During many years there were just two people investigating superstring theories: Green (on the original cartoon he was shown in green colour) and Schwarz (he is black).

Green
Schwarz
O(32)
E8
×
E8

Fig. 5. In 1984 this situation changed, and now there are many people playing strings. Unfortunately, neither supergravity or Kaluza-Klein theory, nor superstrings could solve, without some additional assumptions, the cosmological problems revealed at the end of the 70s.

TOO MANY PROBLEMS...

Fig. 6. One might suspect, that if God knew theoretical physics, he would not even attempt to create the universe. Too many problems: the monopole problem, singularity problem, flatness problem, homogeneity problem, isotropy problem, gravitino problem, ... One should think twice before starting this work!

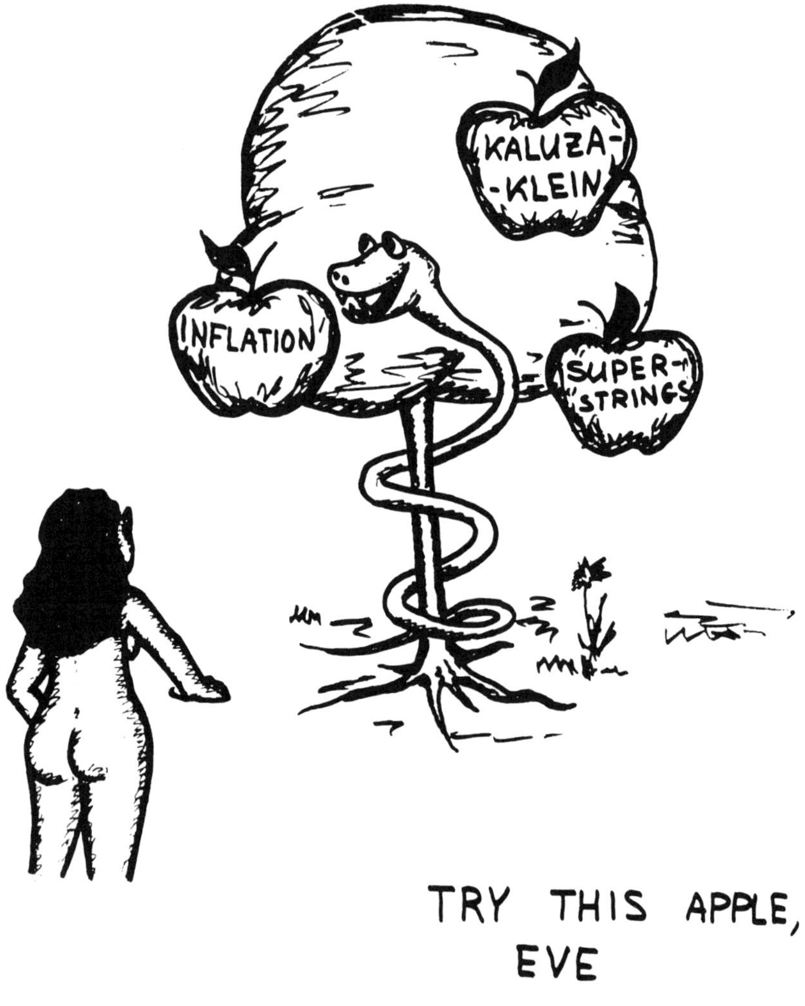
KALUZA-
-KLEIN
INFLATION
SUPER-
STRINGS
TRY THIS APPLE,
EVE

Fig. 7. We do not know for sure who first suggested the idea of inflation; maybe it came about as is shown in Fig. 7. We know, of course, that it is dangerous to eat fruits from the tree of knowledge, but they taste so good!

THE MAIN IDEA OF THE INFLATIONARY UNIVERSE SCENARIO
GRAVITINO
MONOPOLES
DOMAIN WALLS

Fig. 8. There are many different ways to solve one's problems. One possible way is to hide the dirt under the rug. (This is, *e.g.*, what is done in renormalization theory. The great hope associated with strings is to construct a "clean theory," which contains no divergences at all.) Another way is to push the dirt out of one's house. This is the main idea of the inflationary universe scenario. One cannot make the universe totally homogeneous, one cannot destroy monopoles and domain walls, but one can move them outside the observable part of the universe. We know how difficult it is sometimes to keep order in one's own flat. This is why it seems so attractive to have a simple method of doing it in the part of the universe in which we live, - who cares about other parts!

THE OLD RECIPE

Fig. 9. The first versions of the inflationary universe scenario were based on the theory of phase transitions in a hot universe. During the phase transitions some bubbles appeared and expanded exponentially. However, this procedure was very complicated, and the results were not very impressive.

THE PROGRESS IN TECHNOLOGY

Fig. 10. Fortunately, later an advanced technology was invented, based on the idea of chaotic inflation. According to the chaotic inflation scenario, many different exponentially expanding bubbles can be produced from the simplest material available.

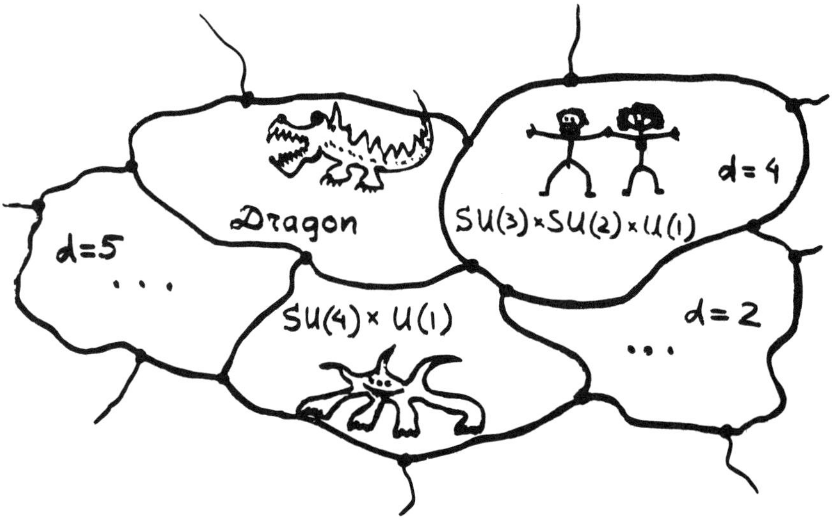

After the compactification the dimensionality of space-time d in different causally unconnected regions of the universe may be different, but <u>life of our type can exist only in d=4</u>

Ehrenfest, 1917

Fig. 11. After inflation the universe becomes divided into many exponentially large domains with different kinds of symmetry breaking (*i.e.* different laws of low-energy particle physics) and even with different dimensionality d of space-time inside each of them. For example, in the supersymmetric $SU(5)$ model there are many different minima of the effective potential $V(\Phi, H)$, of almost equal depth. Some of the regions of the universe after inflation correspond to the minimum with symmetry breaking $SU(3) \times SU(2) \times U(1)$. We happen to live in such a domain. Other domains correspond to the minimum with the symmetry $SU(4) \times U(1)$; we do not know who lives there. Finally, there is a minimum, discovered by N. Dragon, and it is quite clear who can live there.

– I see this bottle to the <u>right</u> of myself,
you see it to the <u>left</u> of myself,
hence, it cannot be real!

– Yes, but its contents
is coordinate-independent...

Fig. 12. The basic principles of the inflationary cosmology are rather simple. However, the large-scale structure of the universe, according to the inflationary scenario, is extremely complicated. Different people studied various properties of the inflationary universe using different coordinate systems. There arose the question of whether the corresponding results are coordinate-independent, which is very desirable if one wants to insist that inflation is something real. This is illustrated by Fig. 12. One man sees the bottle to the right of himself, another man sees it to the left of himself, hence the bottle cannot be real. However, there is definitely something real inside the bottle. Similarly, the most important results of the inflationary cosmology can be formulated in a coordinate-independent way, so that they are also real.

STATUS OF THE INFLATIONARY UNIVERSE SCENARIO SOME YEARS AGO (CAMBRIDGE 1982)

Fig. 13. The first summary of the development of the inflationary universe scenario was given at the conference in Cambridge in 1982. This summary was only partially optimistic: God inflates the universe, but the devil drills a hole into it. At that time it was not quite clear who works more effectively.

Fig. 14. After the invention of chaotic inflation the general picture became much more encouraging. According to this scenario, extremely many inflationary bubbles are produced in different parts of the universe, and the devil does not have the technical means to drill holes inside each of them.

IF IT IS IMPOSSIBLE TO DRILL A HOLE, LET US MAKE IT INHOMOGENEOUS

Fig. 15. Later it was also recognized that it is not so easy for the devil to drill holes in the universe at all. There exist some rigorous theorems, which show that the topology of the universe cannot change outside singularities, at least on the classical level. Maybe, it is possible to do it at the quantum levels, by drilling wormholes through Euclidean space. That is what some people are trying to do now. However, even if the devil cannot easily drill a hole in an inflating balloon, he can make it extremely inhomogeneous. This was the main reason why the first versions of the inflationary universe scenario were unsuccessful.

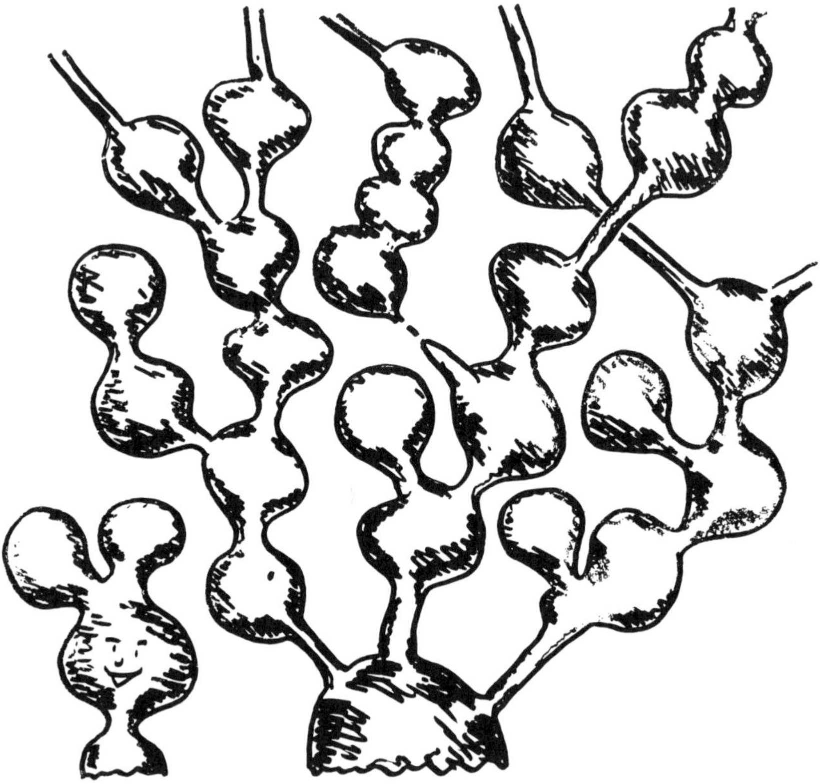

THE RESULT OF THE FRIENDLY
COOPERATION → ETERNALLY
EXISTING SELF-REPRODUCING UNIVERSE

Fig. 16. Now we know how to make the observable part of the universe very homogeneous, but according to most versions of the inflationary universe scenario, the universe is extremely inhomogeneous on a much bigger scale. Surprisingly enough, this distortion of initial harmony of the world proves to be extremely important. If God would work alone, he would create a perfectly homogeneous universe, which would expand and then collapse. However, due to the destructive work of the devil, the universe looks not like a single balloon, but rather as a permanently growing tree, consisting of many exponentially large domains attached to each other. According to this picture, eternity is a result of a friendly collaboration of persons who originally did not have the slightest intention of helping each other.

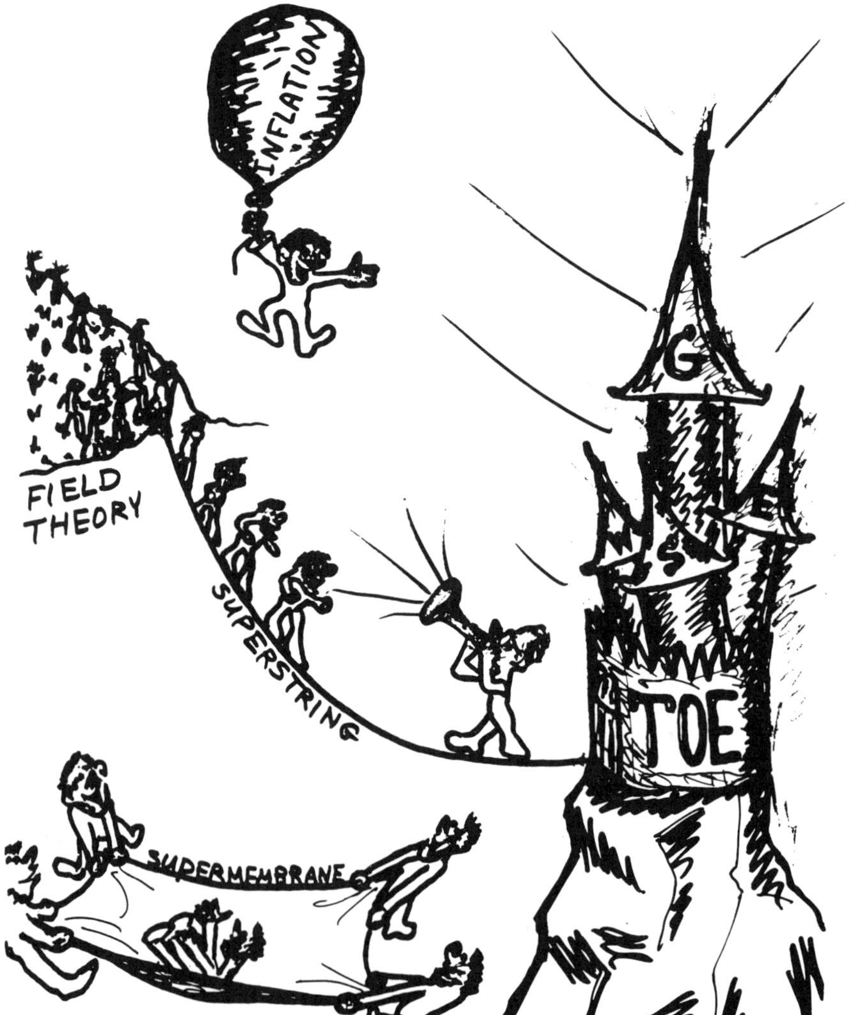
INFLATION
FIELD THEORY
SUPERSTRING
SUPERMEMBRANE
G
E
M
S
TOE

Fig. 17. One may hope that similarly good results may come out of the friendly collaboration between cosmologists and particle physicists. Many people now want to reach the shining castle of the Theory of Everything (TOE), which contains towers of Weak, Strong, Electromagnetic and Gravitational interactions. Previously there was the hope that this could be achieved by simply walking in the Field. However, now many people claim that all grass in the Field is already eaten, and that there is an unbridgeable gulf between the Field Theory and the TOE castle. There is some hope that one can cross the gulf by walking on the Superstring. However, this path is narrow and difficult, and some travellers are falling onto Supermembranes. There is also another way which may help us to come closer to the Theory of Everything, and this is the way discussed in the present book.